Information Spaces

Springer
London
Berlin
Heidelberg
New York
Barcelona
Hong Kong
Milan
Paris
Singapore
Tokyo

Robert M. Colomb

Information Spaces

The Architecture of Cyberspace

Springer

Robert M. Colomb, PhD
School of Information Technology and Electrical Engineering,
The University of Queensland, 4072 Australia

British Library Cataloguing in Publication Data
Colomb, Robert M., 1942–
    Information spaces
    1. Information storage and retrieval systems
    I. Title
    005.7'4
ISBN 1852335505

Library of Congress Cataloging-in-Publication Data
Colomb, Robert M., 1942–
    Information spaces / Robert M. Colomb.
        p. cm.
    Includes bibliographical references and index.
    ISBN 1-85233-550-5 (alk. paper)
        1. Information technology.   2. Information retrieval.   3. Information resources
    management.   I. Title
T58.5.C63 2002
025 -- dc21   2001054256

ISBN 1-85233-550-5 Springer-Verlag London Berlin Heidelberg
a member of BertelsmannSpringer Science+Business Media GmbH
http://www.springer.co.uk

Typesetting: Mac Style Ltd, Scarborough, N. Yorkshire
Printed and bound at the Athenæum Press Ltd., Gateshead, Tyne and Wear
34/3830-543210 Printed on acid-free paper SPIN 10848939

To Thea and the memory of Rosie

Foreword

Bathed in tentative light at the dawn of the Information Age, many people around the world are thrilled with excitement at the vistas just becoming visible. Some envision new information and computer technologies ushering in a Global Information Society where everyone is empowered to realize their full potential. Yet, others grope for ways to relate today's well-understood social institutions with the vague trends only now beginning to take shape. Some societies are being overwhelmed by a deluge of raw information bursting from new media channels, even while most people in the world desperately seek basic telephone service. Information in digital form is decomposed and re-presented so easily that the fundamental premise of a document as a trusted record is under debate. Open societies worry about the potential erosion of the trust relationships on which they ultimately depend. And, virtual communities dispersed across global networks call into question the presumption that physical geography determines boundaries of a society.

Those who work with information in this new reality must understand the nature of information. Knowledge of computer technologies is certainly useful, but sophisticated computer systems often betray sheer ignorance of basic truths long understood in the field of information science. In many cases related in this book, the innovative ideas driving new information infrastructure technology have deep roots in information science. For instance, practical tools for networked information discovery still hold tremendous promise and marketability even after more than a decade of frenzied activity by Internet companies. Yet, practical approaches using a hybrid of human guidance and automated tools are ubiquitous in the "parallel universe" of libraries and information services worldwide. In the Information Age, it may well be that bridging the rift between computer science and information science is the most critical challenge for the academic community generally and for the student of information management specifically.

As the title of this book suggests, spatial and architectural sensibilities can serve as a model to organize a range of important aspects of information. Insights abound when the abstract concepts of information are approached using parallels with spatial constructs and techniques, from "mapping an information space" to "navigating" the Internet. One of the pervasive images is the mind ascending above the problem space to achieve a new perspective. Just as patterns invisible at ground level become clear from a birds-eye view, seemingly disparate areas can be related when so viewed at a higher level of abstraction.

From a broad perspective, for instance, the management of a collection of documents is not an isolated undertaking but connects with the ongoing global enterprise to describe all of the Earth's biology.

Insights into the nature of information are essential whether one wants to make collections easily accessible, to influence public policy or to develop new information technologies. Such insights build steadily in this book, starting from the practical problems confronted in retrieval of information from a collection of documents. Technologies are cited but the focus here is not on choosing a particular solution. The focus is on characterizing the root issues that make each problem worth exploring. As these explorations proceed to build a practical foundation of knowledge and skills, the book notes many fascinating side trips in philosophy and others areas.

The central importance of trust in the context of information management threads through discussions here. Because trustworthiness colors all information content, extreme care is required that information be identifiable always by its source context. A "seamless web of information" may be an attractive technical objective, but information access without careful attention to context only leads to rampant speciousness and confusion. In the final analysis, the most crucial asset of any information collection is the trust established between users and providers. Yet, this goal of preserving trust must be balanced by the goal of open societies to avoid prior constraint on the range of information accessible. It is important to realize that the price of openness is constant vigilance against disruption of the security and reliability of the infrastructure, and against deliberate disinformation.

A practical position is taken here in asserting that typical information collections use textual descriptions. It might be argued that textual descriptions are just one kind of "pattern" within a more general class encompassing many other patterns useful for information retrieval. Examples of fingerprints and faces are noted in the book, and many other pattern-matching technologies are available or arriving soon. In this book, however, the emphasis is on basic information management skills. Text descriptions are certainly the norm for the vast collections of material that are themselves textual or have been catalogued using bibliographic and related cataloging techniques. It should also be noted that most searchers worldwide are either familiar with such text descriptions themselves or are supported by librarians who are, and who assist them in their searching. Even at that future time when generalized pattern matching becomes a ubiquitous set of technologies, the handling of text descriptions will remain a key skill to be mastered.

This book teaches practical approaches to managing information content rather than merely picking among information technologies. Even more important, it teaches that information managers are key players in the global and long-term mandate to make information accessible and usable—information already in cultural treasure-houses worldwide, on the Internet today, and coming at us in ways that ever expand and diversify.

Eliot Christian

Acknowledgements

The Indo-European languages chart which appears as Figure 9.7 is reproduced from the First Edition (Second Revision) of the Macquarie Dictionary (1987) with the kind permission of the publishers, The Macquarie Library Pty Ltd, North Ryde.

Figure 12.3 is reproduced with the permission of Andrew Smith, Leximancer, August 2001.

Figures 13.9 and 13.10 are reproduced with the permission of David Nation.

Contents

1 Introduction

With the development of the World Wide Web, cheap CD-ROMs and, generally, easy access to computing and communication technologies, it has become easy to publish collections of documents, so that a huge variety are now accessible to a wide audience. Some of this variety is shown in Table 1.1.

Table 1.1 Some kinds of collections of documents.

- Books – both libraries and bookstores
- Articles – newspapers and magazines
- Memos and e-mail messages in large organizations
- Abstracts – a wide range of bibliographic databases
- Legal cases and statutes in many jurisdictions
- Web sites – organized by search engines collections of databases
- Images
- Museum collection catalogues
- Music fragments
- Criminal information, both on-going police records and specific investigations, including mug shots and fingerprints
- Records/CDs
- Videos, movies
- TV news footage
- Software libraries
- Queries, problems, frequently asked questions for software products
- Submissions to public enquiries
- Telephone hotline calls for product contamination incidents

People sometimes use these collections to find particular documents according to various criteria, and sometimes want to find out general characteristics of the collection, either as some form of statistical content analysis or some type of visualization. Publishing the documents therefore includes provision of access to information retrieval, analysis and visualization technology of various kinds.

Provision of technology is easy and it makes life easier for the users of the document collections, but its utility is limited if it is used on the raw document collection. The problems are not deficiencies in the technology, but are inherent in the way the documents are created and searched. At present, nearly all documents are texts in natural language. Even collections of images are generally represented by catalogue entries and captions, which are searched instead of the image itself. Different documents are created by different people for different purposes, so the vocabulary used and the assumed context differ from document to document.

Further, the person searching the collection generally has a different purpose in mind than any of the document creators.

Of course, even though many people and organizations are publishing documents for the first time, the problem of managing and organizing large collections of documents is not new. Librarians have been managing public collections for hundreds of years, and commercial bibliographic databases have been available for decades. These collections have historically been managed by information science professionals of various kinds, including librarians, records managers and information managers. These people make their collections easier to use by introducing intermediate information structures in addition to the raw text of the documents. These structures broadly include standardized subject and keyword descriptors and classification systems, together with standardized methods of cataloguing the documents.

Successful publication of collections of documents therefore requires provision of additional intermediate information structures to facilitate the user's task of finding things and getting overviews of the collection. In most cases, it is not economic to employ a professional librarian just as it may not be economic to employ a professional software engineer. The collection is typically managed by someone who understands the domain, with sufficient information technology expertise to operate and configure the software products used. In the same way, the manager of the collection needs to know the basic principles of information science.

This book is aimed at the managers of document collections, and those university students who expect in their professional life to have occasion to manage such collections, without being professional librarians. It is organized in two main parts – the first oriented towards the problem of finding documents in the collection and the second towards the classification systems and controlled vocabularies which not only assist in information retrieval but also make possible an overall view of the collection. It begins by considering the problem as the retrieval of information from a collection of documents, but by the end the problem is seen as the user navigating in an information space.

The information retrieval part opens with an overview outlining the main issues, then proceeds to a discussion of the technologies employed for retrieval and how they are used, then to the information structures used to support information retrieval. Chapter 4 turns to the use of hypertext and the strategy of browsing rather than searching, then to the World Wide Web and the problem of resource discovery in that vast and heterogeneous environment. The information retrieval part concludes with a consideration of documents which are themselves complex structures and the use of the markup language XML to represent and exploit that structure.

Until this point, the text has taken the perspective of the user trying to find information in the document collection. The second part takes the perspective of the managers responsible for the information structures used to organize the collection. It begins with a discussion of classification systems and their basic design principles, then how these classification systems can be used to organize and present the collection. Chapter 9 makes the point that classification systems are designed, not inherent qualities of the documents. This leads to the main design principles, first of classification systems then of subject and keyword systems. The structure part concludes with the use of the structures to support sophisticated visualization of the information spaces.

Two final chapters look at the main issues in creating and managing archives, which are the ultimate source of many published document collections; and general issues of quality, including legal considerations.

The book is intended to assist the reader to learn a specialized technical vocabulary used to describe information-seeking behaviour, the design elements of information spaces and the metrics used in the evaluation of design choices. In particular, the reader will learn to use these concepts in design and criticism of design of information spaces. There are about 75 key concepts, which are gathered into a glossary at the end.

Use of this book as a text is facilitated by exercises and discussion questions at the end of each chapter, and by the inclusion of a sample set of assignments which have been successfully used in teaching this material to a broad group of second-year students at The University of Queensland.

As a university text, this book is relevant to a wide range of disciplines, including journalism, languages, anthropology, management, public administration, law, education, social work, and engineering, as well as more application-oriented information technology students. It is relevant to health informatics, tourism informatics and biological informatics – to any field of activity which generates or requires access to collections of documents.

2 Overview

This chapter gives an overview of the issues in the text. What do we mean by information? What are the main issues in its storage and retrieval? How are the information spaces organized on a large scale? It introduces text databases.

What is information and what can be done with it?

In order to act effectively in a situation, a person or an organization needs to know the relevant aspects of the situation. For example:

- In order to choose a course of tertiary study, a person needs to know what their interests and objectives are, what institutions have courses relevant to those interests and objectives, how these courses work, their entrance requirements, procedures for application, the reputation of the courses and institutions, possibly trends in the field and how the courses are adapting to them, and so on.
- A company wishes to locate a new retail outlet. It needs to know possible sites, and for each site the number of people with particular characteristics in the area. It may also need estimates of traffic through the area, and the location of other relevant retailers in each area. Further, it could be valuable to know about trends in the area and development plans or proposals.
- A development is proposed for a particular area, and a government agency must assess its potential impact on the area in order to decide whether to approve it. This requires knowing what existing land uses are and how each might be affected, what species of flora and fauna might be affected, and what other populations of affected species exist in the region. In addition, the opinion of interested parties might be considered as submissions to a public enquiry.

Much of what needs to be known is represented in documents of one kind or another (some of the variety of possible documents were listed in Table 1.1). Information science is concerned with collections of documents as they are potentially relevant to particular purposes.

More specifically, information science is concerned with tools which can assist people in their deliberations by helping to extract relevant information contained in document collections. These tools fall into two basic categories, corresponding to the two basic things one can do with documents in a collection: find individual documents, and organize the collection to get a total view.

Find a document

The process of finding a document (called *search*) depends a great deal on how much we know about the document we are looking for. Sometimes, we can identify a particular document. For example:

- the library call number and borrowing status of a book given its title;
- the price and availability of a book given its ISBN;
- the Jeremy Horey column in *The Australian* of Tuesday, 23 February, 1999;
- the article by Clifford Lynch in the reading list for the course CS270 for 1999.

However, more often we are looking for information without knowing what documents contain what we want, or even if the collection even contains any documents we want. For example:

- books in the library on information science;
- any books either in the library or in print by Raymond Smullyan;
- articles in the information technology trade press about Web portals;
- reputable information on antibiotic-resistant microbes and their presence in hospitals.

Many of the collections of documents listed in Table 1.1 are not textual, for example mug shots (highly regular photographs of a person's face, often used in criminal records), or collections of stock photographs held by libraries or archives of various kinds. Here we are often looking for documents that match certain non-textual characteristics, for example:

- mug shots of people matching a description of an offender given by a witness as a police identikit picture;
- photos showing the fashions worn by upper-class women in the 1920s in Baltimore.

Ideally, one would like to be able to retrieve these documents by an automatic search of their contents. In fact this is possible in some cases. In particular, for mug shots there are at least two methods.

- An identikit consists of a number of facial features (hair, nose, eyes, mouth, chin, etc.), and for each feature a number of variants. The witness constructs the description by choosing the closest variant for each feature. The library of photographs is preprocessed, and to each is attached the identifier of the closest variant for each feature. This can be done using image processing software. The software can then match the description's variants with the stored variants.

- There are certain characteristic features of a person's face that are stable over long periods, including distance between the eyes, distance between nose and chin, ratio of width to length of head, and so on. Each mug shot is processed to extract these features using image processing software. The witness's description is converted to an image by either identikit or an artist's sketch, which is also processed and its features extracted. The retrieval software can match the description's features with the stored features.

Methods like this work in highly constrained fields, notably fingerprints, but in general, an image is given an associated caption, which describes its contents in words, and the image collections are searched by text queries on the captions, in exactly the same way as textual documents.

Organize a collection

Finding a document or a small set of documents in a collection is one common activity. The other main thing people want to do with a collection of documents is to get an overview of the whole collection. For example:

- How many documents are in the collection?
- Dividing the collection into types, how many documents are there of each type?
- In a collection of newspaper articles, what are the 100 most common themes?
- In a collection of unstructured interviews, what topics are discussed, and how often in each?

Another way to look at the whole collection is to consider that some documents are in some sense nearer to each other than others. Often what is desired is a nearness measure based on semantic content: two books are similar to the extent that they cover a similar range of topics. Alternatively, the nearness might be on structural grounds: in a collection of television programmes, one police show is very close to another, a little further from a private detective show, a little further again from an action thriller, further again from a western, and a long way from news programmes, cooking shows and travelogues.

One might want to browse the collection using the nearness relationship, starting at one document and being able to traverse to nearby ones, perhaps eventually reaching a document quite distant from the starting point. Starting from a book on information science, one might get successively a book on text databases, a book on structured (SQL) databases, a book on information systems development, then a book on the Battle of Britain (because the book on information systems development used the organization of the flow of radar and planespotter information in that battle as an extended example).

Alternatively, one might want to use the nearness relationship to get an overview of the whole space. Starting from a particular police show, one might want the names of other shows very close to it (other police shows), then an indication of detective shows without names, then progressively less distinct indications of further shows. This sort of picture can be represented as what is called a fish-eye view, as discussed in more detail in Chapter 9.

The most common way to browse a collection of documents is via hypertext. In hypertext, each document contains links to other documents. Hypertext is viewed with a piece of software called a browser, which traverses a link by following a command by the user. Two documents connected by a hypertext link can be thought of as nearer to each other than two documents that are not connected. Hypertext links are in principle extremely specific – this particular document is linked to that particular document – but links are often grouped into types. Hypertext is discussed in more depth in Chapter 5 and in the context of the World Wide Web in Chapter 6.

Information structures supporting the collection

Finding documents in a collection and getting an overview of a collection are significant issues for other than specialist librarians because it is possible to use computer systems to help in these tasks. The first issue to be considered is therefore how the document collections can be represented in the computer.

Using the examples in Table 1.1, some collections are easy – collections of e-mail messages for example are already computerized objects. Other collections, notably books, are of objects which are physical, and are not stored in the system in and of themselves. Books must be represented in some computerized form before computers can be used to help find them or to help overview the collection. This is typically done with a catalogue entry, which is generally a small piece of text which represents the book, and which contains a description of the book. Other collections are of objects which may be stored in the computer but which cannot easily be searched by computerized means – photograph archives for example, which as described above must be represented by a caption in the computer system. Even documents which are stored in computer systems, such as Web sites or reports, are often represented in information storage and retrieval systems by abstracts and other brief descriptions.

In this text we use the blanket term "surrogate" for a document which stands for some other document in an information storage and retrieval system. A *surrogate* will generally contain some *descriptors*, which are terms that concisely describe the content. The whole system is pictured in Figure 2.1. A user has some *information need* which can be filled by a selection of documents from the collection. The *information storage and retrieval system* mediates between the information need and the documents – the user formulates a *query*, the system searches the descriptors in the surrogates and returns a *response* consisting of a set of surrogates. The user examines the surrogates, making a judgment as to whether any

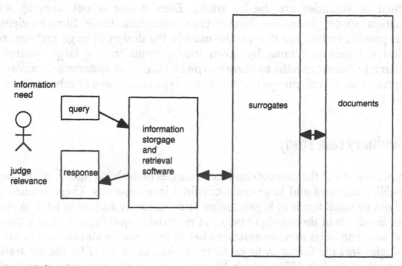

Figure 2.1 An information storage and retrieval system mediates between a user's information need and a collection of documents.

of the documents referred to are likely to satisfy their information need. Chapter 4 looks at the design issues involved in creating surrogates.

The information storage and retrieval system can have a very simple technology – simply performing a high-speed exhaustive search of all the surrogates looking for matches to the query. Often, however, the collection of documents is large, and this very simple solution is not fast enough. Even searching at 1000 records per second, it takes 10,000 seconds or nearly three hours to search 10 million surrogate records. As we will see in Chapter 3, more sophisticated software solutions are available for larger problems.

Some documents are highly structured, often large – for example the text of a book marked up for typesetting, or an electronic data interchange (EDI) document. It is possible to represent the structure of these documents in a standard way that can be useful in searching or processing its contents as well as controlling its presentation. Chapter 7 looks at the XML Internet standard markup language from this point of view.

Classification systems and other controlled vocabularies are the main structures supporting the large-scale view of document collections. This is how collections are divided into types, how themes are identified in newspaper articles, and topics in unstructured interviews. General problems in the design of such systems are discussed in Chapter 8. Classification systems can be used to define the geometry or topology for visualizing the information space – visualization is becoming more and more important with the increasing power and decreasing cost of personal workstations. Useful properties of classification systems and design issues relevant to visualization are considered in Chapter 9 and their application to visualization in Chapter 13.

Despite the ubiquity of classification systems and their centrality to information spaces, classification systems are not inherent properties of the data but are designed objects suitable for particular purposes. How this is, and some alternative ways of achieving particular goals, is the subject of Chapter 10.

One of the major professions which have historically had charge of large controlled vocabularies are the librarians. Even if one is not working with information spaces based on library-type materials, these library-oriented systems provide techniques that can be used in the design of large systems, and also design issues that must be taken into account in any large controlled vocabulary. Problems specific to library-type classification systems are canvassed in Chapter 11 and problems specific to library-type collections of subject terms in Chapter 12.

A preliminary case study

One important use of the concepts and technologies underlying information spaces is in public enquiries and large-scale criminal investigations. These studies are carried out by small teams of highly trained professionals, and often have to make sense of hundreds of thousands of pages of material to quite specific ends. One of the first such enquiries to make extensive use of information storage and retrieval technologies was the Costigan Royal Commission, established by the Australian government in the late 1970s, which began as an investigation into a particular trade union, but which uncovered large-scale organized corporate fraud and tax evasion. (A Royal Commission is an investigation into a specific issue conducted by a judge, with powers to compel witnesses and other legal powers.) It is common in Australian Royal Commissions for a prominent barrister to be appointed to lead the team of investigators, in this case Douglas Meagher, QC. Because the use of information technology was novel and very useful to the investigation, Meagher described how the technology was used in a number of speeches, providing the material for the case study below.

The problem of investigating organized crime is stated by Meagher as

Although occasionally the identity of the criminal organization will be notorious, more often than not its existence is not known or is the subject of merely vague rumor. In these cases the only information available is a large number of isolated occurences of crime. A very high proportion of those will have been committed as individual acts, with no relation to an organization. The investigator is required to survey the field, to sift out those which are isolated and not part of an organizational pattern, and then examine the remainder with a view to identifying the organization concerned. This task is made more difficult by the circumstance that several criminal organizations are at work, some known, some not. The investigator has to endeavor to analyze each of the matters to determine where it properly belongs.

A large amount of material must be amassed in order to be sifted. The process begins with collection of material of the most disparate sort – arrest records, witness statements, documents such as diaries and address books seized in search of properties, telephone, company and financial records, statements of informers, correspondence, and so on. Since the significance of any piece of information may not be clear until correlated with many others, the collection process uses only the most rudimentary assessment of significance. Everything goes in.

Information collected is not, in its raw form, very useful. For one thing, it exists in a variety of media not all of which are easily made computer-readable. For another, the pieces of information have all been taken out of context, so it can be hard to see the significance of their specific content. A diary entry saying "John Walker 3pm" on the page for Tuesday, 6 April 1999, is unintelligible unless it is also recorded that the diary belongs to James Beam and was found in a search of the premises of "White Horse, Inc." during a search on a given date for material related to a suspected tax evasion event. All material, whether machine-readable or not, must therefore be represented to the information system by registration.

Registration involves making a catalogue entry which includes the person or organization involved, the date and circumstances of its collection, the type of material, and also key words from a standardized collection of terms which describes its contents. The diary entry of the previous paragraph might have its content described by the keywords "diary entry", "John Walker", and "6 April 1999".

Registered information can now be retrieved, but only by the most elementary criteria. To maximize its usefulness, each new item must be collated. Indexes of people, events and investigations are constructed during the course of the investigation, and each registered item must be cross-referenced with these indexes. It might be known that the John Walker of our diary entry is a known alias for Ezra Brooks. The diary might have been seized during an investigation of the collapse of a group of companies associated with James Bond.

Furthermore, particular types of information have other keywords and codes attached to them during this process. People have descriptions coded, telephone numbers are identified, and financial information is recorded in standardized detail – checks, deposit slips, bank statements, receipts – which are totalled by type and time period.

Collation is performed by trained data entry clerks assisted by specialists for less routine entries, the whole supported by indexes and dictionaries of codes, people, organizations, investigations, special data entry forms for specific kinds of information, etc.

Finally, the analysis. The whole enterprise is intended to make the most effective use of the small number of highly skilled analysts available to the investigation. Analysis is supported by a range of tools, including simple retrieval of records associated with particular people or organizations, retrieval of all records meeting defined characteristics from the various codes and keywords, the ability to sort sets of retrieved records by date and type, the ability to compare the sets of records from one query with the set from another – identifying like and unlike data (useful in identifying false names, for example) – and the ability to identify networks of links among the objects of investigation. James Beam is connected to Ezra Brooks by the diary entry (Ezra Brooks is the real name of John Walker). Brooks is recorded as having paid money to Jacques Daniels, who is a director of one of the companies of which James Bond is a co-director. There is therefore an indirect link between Beam and Bond consisting of three indirect links.

Meagher says that the analytical tools provided by the information technology were of enormous value in making sense of the approximately 40 million pieces of information collected by the Costigan Royal Commission. However, and most importantly, he points out that the results of the analysis of the collected, registered and collated data are never definitive. He says that links and patterns discovered by the analysis are only hypotheses, not fact. To proceed with criminal charges these hypotheses must be investigated by the police, resulting in further information which may serve to confirm, disconfirm or change the hypotheses.

This case study is but one of the many possible uses to which the concepts and technology of information spaces may be put.

Key concepts

 An **information storage and retrieval system (ISAR)** is used to organize and access a collection of **documents**. An ISAR system allows the user to either **browse** the documents or **search** for documents with particular characteristics. A document is represented in the ISAR system by a **catalogue** entry. The catalogue entry is a **surrogate** for the document, containing **descriptors**.

3 Text retrieval

This chapter begins the "find a document" part of the text. It covers information storage and retrieval systems, query languages, and retrieval strategies.

Query mechanics

A computer system to search a collection of documents is called an information storage and retrieval (ISAR) system. These systems are conceptually quite simple. Each document contains a number of terms, and the system has a basic query: *Find X* means "return all documents containing term X".

A term can be a single word: "information", "storage", "retrieval", "system", "ISAR"; or it can be a phrase "information storage and retrieval system" or "information retrieval". Query processing is based on character string matching. The query processor behaves as if it matches the character string in the query with the character strings in each document, returning the identifier of each document containing a string matching the query string. It is important to remember that this matching has no intelligence. If the query string is "system" followed by an end-of-word mark and a document contains the string "systems" followed by an end-of-word mark, then the strings will not match, and the document containing "4 systems" will not be retrieved. (The query processing can be improved, as we will see below, but essentially simple string matching is what is going on.)

A common type of query language permits several basic queries combined using the boolean connectors "and", "or" and "not". If two queries are connected by "and", then the result is the set of documents contained in both query result sets; if "or" then the result is the set of documents contained in either result set. *Find "information and retrieval"* is the result of *Find "information"* intersected with the result of *Find "retrieval"*, that is, the set of documents which contain both terms. *Find "information or retrieval"* is the result of *Find "information"* union the result of *Find "retrieval"*, that is, the set of documents which contain either term.

The negative connector "not" is generally used together with "and". If two queries are connected by "and not", then the result is the set of document containing the first term but not the second. *Find "information and not retrieval"* is the result of *Find "information"* with any document in the result of *Find "retrieval"* removed.

Query languages using the boolean connectors are called *Boolean* query languages. They are very common in bibliographic systems.

The problem is not the query

We have a query processing system that can retrieve documents containing or excluding particular combinations of words. However, in the discussion of the problem of finding information illustrated in Figure 2.1, the query is only the second step in the process – it begins with a user's information need, as highlighted in Figure 3.1.

Use of an ISAR system to find information is considered to have four steps:

1. A user has a need for some information for some purpose, say, finding the current state of database technology for managing criminal investigations like the Costigan Royal Commission discussed in Chapter 2.

2. The user constructs a query requesting the ISAR system to retrieve documents containing particular combinations of terms – say, Find "database and technology and criminal and investigation".

3. The ISAR system returns a collection of documents.

4. The user examines the documents returned, assessing them for relevance to their information need.

What will typically happen is that the result set will contain some documents that fail to meet the information need. The query of Step 2 would retrieve a document containing the phrase "investigation of the vulnerability of database technology to criminal fraud". This is a case of a retrieved document that is not relevant. Furthermore, there will very likely be documents in the collection which would meet the information need, but which are not selected by the query for the result set. Note that the discussion of the Costigan Royal Commission in Chapter 2 does not contain the term "database", so would not be in the result set. This is a case of a relevant document not retrieved. However, with luck, some of the documents retrieved will also be relevant, so that the user will get some information that can be used. The situation is illustrated in Figure 3.2.

Because user need and judgment are factored into the design of ISAR systems, the problem of irrelevant documents retrieved and relevant documents not retrieved is central to their evaluation. Consequently, these aspects of performance are measured using two technical terms, *precision* and *recall*.

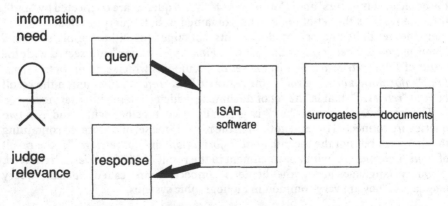

Figure 3.1 A user's interaction with an ISAR system.

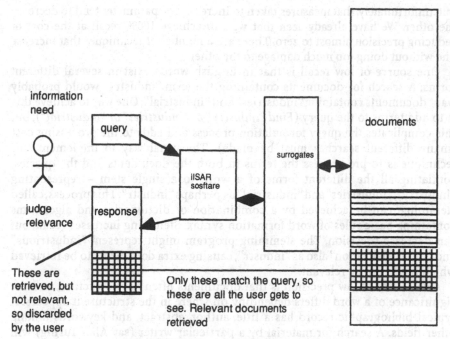

Figure 3.2 The performance of an ISAR system..

The precision of the response to a query is the percentage of retrieved documents relevant. In Figure 3.2, it is the cross-hatched area in the response box as a percentage of the total area of the box. If the response contains 20 documents and 6 of them are judged relevant to the user's information need, then the precision of that response is 30%.

Recall is the percentage of relevant documents retrieved. In Figure 3.2, it is the cross-hatched area in the documents box as a percentage of the area highlighted. If the collection contains 18 documents relevant to the user's query and the query returns 6, then the recall is 33%. (Note that the cross-hatched area in the documents box and the response box represents the same documents, the relevant documents retrieved.)

It is, of course, much easier to measure precision than recall, since precision can be calculated from the set of documents returned by the ISAR system as the response to a query, while to measure recall one would need to know all the relevant documents beforehand, which eliminates the need for the query at all.

There is an easy way to get 100% recall, though – simply put all documents in the response (*Find all documents*). If the user examines all the documents in the collection, then they will necessarily find all the relevant ones. This easy solution is, of course, not practical. ISAR systems exist because an exhaustive search of a large collection of documents is far too expensive if done manually. Designers of ISAR systems can estimate recall using various methods, but for a particular user to estimate recall for a particular search is a serious issue, which will be considered below.

Precision and recall are design parameters for ISAR systems. The designers would like to achieve systems where both parameters are as high as possible. It turns out, unfortunately, that measures taken to increase one parameter tend to decrease the other. We have already seen that we can achieve 100% recall at the cost of reducing precision almost to zero. There are a number of techniques that increase one without doing too much damage to the other.

One source of low recall is that in English words exist in several different forms. A search for documents containing the term "industry" would probably want documents containing "industries" and "industrial". One way to achieve this is to add terms to the query (*Find "industry" or "industries" or "industrial"*), but this complicates the query formulation process and adds to the processing cost (many different searches must be made). The usual way to implement this technique is to pre-process the terms in both the documents and the queries, conflating all the different forms of a word to a single stem – representing "industry", "industries" and "industrial" as perhaps "industr". This process, called stemming, can be achieved by a combination of dictionaries and algorithms containing basic rules of word formation syntax. Stemming increases recall, but can decrease precision. The stemming program might represent "industrious" and "industrialization" also as "industr", causing extra documents to be retrieved which might not be relevant.

One source of low precision is that documents often have structure, and the significance of a word differs depending on where in the structure it appears. A typical bibliographic record has a title, author, abstract, and keywords among other fields. A search for material by a particular writer (say Alan Turing) will return books by Alan Turing, books about Alan Turing, and books whose approach to material is said to be similar to Alan Turing's. If the ISAR system allows the place in the structure to be specified, the user can restrict the query to the author field (*Find author = Alan Turing*), eliminating the other two categories, thus raising precision. Recall can suffer, however – restriction to the author field could miss edited collections of articles containing significant papers by Alan Turing.

Another source of low precision is that combinations of words may occur fortuitously in a document. A query *Find "information" and "retrieval"* would find a document containing the phrase "information given to the police concerning suspicious behaviour of people making their living by retrieval of bodies from the River Thames" (in Dickens' *Mysterious Stranger* say). Some ISAR systems allow the user to specify that two terms be adjacent in the same sentence in the document. This restriction would eliminate the *Mysterious Stranger* document, increasing precision, but would miss a document containing the phrase "IBM's STAIRS product was one of the first database systems used for large-scale retrieval of information", thereby reducing recall.

There are many other simple sources of low precision and recall, including abbreviations, US/UK spelling conventions, and differing transliterations of names from non-Latin alphabet languages. Furthermore measures taken to improve recall and precision generally have the reverse effect when applied to negated elementary queries. For example stemming "industry" to "industr" includes more documents when applied to a positive elementary query, but excludes more documents when applied to a negated elementary query.

These cases could be multiplied, but such simple situations are not the fundamental problem. Getting high precision and recall is *hard*.

Why is it hard to get high precision and recall?

We begin by illustrating the problem. Suppose we wish to find information about design of data in database systems, with an emphasis on the re-use of designs. A plausible query to obtain such information might contain the keywords:

Query 1: programmer and database and library and compile and record

We would be satisfied with a document containing the following text, which contains all five of the keywords in the query.

Document 1:

A **programmer** developing a **database** will identify the information needs of the client. Data needs will be expressed as specifications for data **records**. Ideally, the organization's **library** of existing specifications will contribute a significant portion of the design, which will then be **compiled** into a prototype.

This response is a *match* to the query (a relevant document retrieved). There might be another document containing the following text:

Document 2:

A knowledge engineer developing a rule-based system will identify the knowledge needs of the client. The knowledge acquired will be expressed as the specifications for slots and frames, and their relationships. Ideally, the organization's archive of existing knowledge will contribute a significant portion of the design, which will then be assembled into a prototype.

This text addresses a very similar topic, and which might be considered relevant to our problem. This document contains none of the keywords in the query, however, and would not be retrieved. It contains none of the keywords since the field of knowledge-based systems originated in the field of artificial intelligence (AI), which was separated from mainstream computing in the 1960s. When AI began to produce industrially relevant technology in the 1980s, a distinct vocabulary was developed to describe activities very similar to those performed by information systems professionals, and to describe closely related technology. This second document would be a *miss* or a *false negative* with respect to the query, since it is relevant but not retrieved.

Finally, consider the following document.

Document 3:

The **programmer** for "music on a theme" consults the **database** for our music **library**, and **compiles** a selection of **records**, which will make a pleasing program.

This document is from a completely different field, so is not relevant to the query. It is retrieved because it contains all five of the keywords in the query. The database system does not know that the words are used in a very different sense in the document from the intention of the person making the query. This is an example of a *false match* or a *false positive* (retrieved document not relevant).

Consider, however, that the person making the query might have been interested not in computing, but in the operation of music radio stations. This second person might have formulated the same query,

Query 2: programmer and database and library and compile and record

and have wanted the third response. To this person, the first response would be a false match.

How do we tell that Document 1 is a desired response to Query 1, while Document 3 is not? Similarly, how do we tell that Document 3 is a desired response to Query 2, while Document 1 is not? We might say that Document 1 is about computing while Document 3 is about operating a radio station. What evidence do we use to make these judgments? After all, the term "computing" does not appear in the text of Document 1, nor does the term "radio" appear in Document 3.

The process of understanding texts is very poorly understood, but a plausible process might be:

1. We look at the words in the text, and how they are organized by the grammar of the language.
2. We are familiar with the sorts of words and their grammatical organizations used in computing and in radio broadcasting.
3. Using the evidence of the text, we are able to classify the texts into the categories with which we are familiar.
4. We look further at texts which are in categories relevant to our interest, and not at texts which are not in relevant categories.

Steps 2 and 3 in the process rely on information which is not strictly contained in the texts. In fact, the user must have considerable knowledge of the topic in order to make any sense of the responses. It is very difficult to search on a topic of which one is totally ignorant.

This knowledge is not generally conveyed to the ISAR system. The texts of Queries 1 and 2 are identical. There is no indication whatsoever that the person making Query 1 is interested in computing and not radio broadcasting, while the person making Query 2 has the reverse interests.

An information retrieval system is a computer program, which is only able to perform computations on the data available to it. **It is therefore unreasonable to expect that the system will be able to retrieve all and only the texts relevant to a query, since the system simply does not have enough data available to it.** There are many who think that it is in principle impossible for an information retrieval system to have enough information available to it to be able to retrieve all and only the relevant information. The arguments about this are beyond the scope of this text, but at least all agree that it is possible, albeit extremely difficult, to solve the problem.

A text-based information retrieval system is therefore a tool to assist people in finding useful information. It has only the text of the documents in its database together with the text of the query expression created by the user to work with. As we will see later in the book, it may have additional structures such as a dictionary or thesaurus to help in its task, but these are also words. **An information retrieval system works only with words, not with their meaning. It cannot therefore give exact results if its performance is judged on matching meanings.**

We must therefore expect limited precision and recall.

Document nearness query languages

We began this chapter by considering the basic query *Find X* and how several basic queries can be combined in a boolean query language. A second type of query language uses the basic query in a different way.

There is an intuition that some documents are nearer to each other in meaning than others. For example, a newspaper article on a tornado in Shreveport, Louisiana might be considered very similar to a newspaper article on a tornado in Lubbock, Texas; less similar to a newspaper report of a cyclone in Vanuatu; and very different from a newspaper report of a business collapse in Italy. The extent to which this intuition holds up under investigation is a question, but not a question for this book. However, the intuition that some documents are nearer to each other than others has led to an information retrieval technology that is in wide use, especially in Web search engines.

Recall from the previous section that ISAR systems do not have access to the meaning of documents, only to the words making them up and to auxiliary collections of words. Therefore, in practice the measure of nearness of two documents is the proportion of words they have in common. This measure would range from 100% for the nearest to 0% for the most distant. Clearly, a document is as close as possible to a copy of itself. The highest value of nearness will be 100%, reflecting that all words are in common.

Now let us consider two documents that should be as far apart as possible, that is a nearness of 0%. The first is part of the quotation from Meagher from Chapter 2.

Document 4:

Although occasionally the identity of the criminal organization will be notorious, more often than not its existence is not known or is the subject of merely vague rumor.

And the other is part of a nonsense poem by Edward Lear "The Owl and the Pussycat".

Document 5:

The owl and the pussycat went to sea in a beautiful pea-green boat.

An inspection reveals that Documents 4 and 5 have only one word in common, "the" which occurs three times in Document 4 and twice in Document 5. The co-occurrence of "the" in two documents seems pretty far-fetched as evidence that the two documents are somewhat similar. Nearly every English-language document contains words like "the", "a", "in" etc. For this reason, document nearness measures generally exclude these common words by keeping a *stop list* of such words (called *stop words*), and removing them from the documents before computing nearness. Documents 5 and 4 might therefore be reduced to:

Document 4':

occasionally identity criminal organization notorious, more often existence known subject merely vague rumor.

and

Document 5':

owl pussycat went sea beautiful pea-green boat.

which have no words in common, and therefore a 0% nearness.

Between these two extremes, the nearness measurement is somewhat arbitrary and is defined differently by different systems. One simple way is to take the number of words in common as a percentage of the number of words in the smaller document. More elaborate systems take into account the number of times the word occurs in the document, or will weight some words as more representative of content than others. This kind of issue is discussed in the next chapter.

Document nearness is used as the basis of a query language by considering the query as a document. The ISAR system then returns documents in order of nearness to the query.

Note that what is called here "nearness" has many other names in the literature, including "document vector approach", "best match", "relevance ranking" and "document distance". The term "nearness" has been employed as it fits better with the text's overall theme of information spaces.

We can compute the document nearness between Documents 1, 2 and 3 based on the number of shared terms divided by the number of terms in the smaller document, as in Figure 3.3.

Notice that Document 2 is nearer to Document 1 than is Document 3. That is, if a document nearness query language were used and Document 1 used as the query, then Document 2 would have been retrieved ahead of Document 3. This is a very different result from the boolean query above, where Document 2 was missed altogether. One might speculate that this result illustrates a superiority of nearness languages over boolean languages.

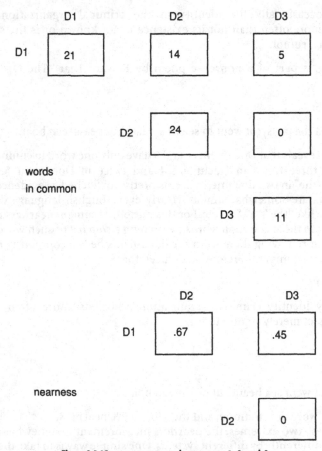

Figure 3.3 Nearness among documents 1, 2 and 3.

To investigate this possibility, we examine the two documents. Even though the keywords from Query 1 are not in Document 2, the two share words like *client, contribute, design, developing, existing, expressed, need, organization, portion, prototype, significant,* and *specification.* These words have little to do with information technology, but much to do with the process of design. So a document about the design of power plants or even an advertising campaign would contain similar terms, and would similarly be close to Document 1.

This example shows that the nearness-based query languages only produce different result sets, but do not change the problems of precision and recall.

Index construction

Search operations in boolean or document nearness query languages are generally carried out on simple data structures called *indexes.* An index is produced from the text of the documents in the collection in more or less the following manner.

First, each document is processed to identify the distinct words it contains and how many times each occurs. This process, applied to Documents 1, 2 and 3 above, is shown in Figure 3.4. The headings D1, D2 and D3 refer to Documents 1, 2 and 3 respectively.

These word lists derived from the different documents are then merged, as in Figure 3.5. Each word is now annotated with the document it comes from as well as the number of times the word occurs in that document.

Finally, the stop words are removed and the remaining words are stemmed. Attached to each term is now the list of documents in which it occurs., together with the number of times it occurs in that document, as in Figure 3.6.

Boolean queries are computed from the index simply by performing the set equivalents of the boolean operations on the lists of documents attached to each of the query terms. Document nearness-based query languages are computed from the index by merging the document lists associated with each of the query terms, then sorting the result by document, counting the number of terms associated with each document and finally sorting again by the number of query words associated with each document. Various weighting operations are sometimes performed.

How does a boolean ISAR compare with a nearness-based ISAR?

The difference between boolean and nearness-based query systems can be seen most easily by example. Take Query 1 "programmer and database and library and compile and record", which contains five terms. A boolean ISAR will return all documents containing all five terms and no document not containing all five terms. The corresponding query in a nearness ISAR would be Query 1 with the boolean connectors removed:

Query 3: programmer database library compile record

The nearness ISAR would first return all documents containing all five terms, so the first part of the response would be the same as for the boolean ISAR. However, it would then return all documents containing four of the terms (any four), then all

D1	D2	D3
a 4	a 4	a 3
as	acquired	and
be 2	and 2	compiles
client.	archive	consults
compiled	as	database
contribute	assembled	for 2
data 2	be 2	library,
database	client	make
design	contribute	music
developing	design	of
existing	developing	on
expressed	engineer	our
for	existing	pleasing
ideally	expressed	program
identify	for	programmer
information	frames	records
into	ideally	selection
library	identify	the 2
needs 2	into	theme
of 3	knowledge 4	which
organization's	needs	will
portion	of 3	
programmer	organization's	
prototype	portion	
records	prototype	
significant	relationships	
specifications 2	rule-based	
the 4	significant	
then	slots	
which	specifications	
will 4	system	
	the 6	
	their	
	then	
	which	
	will 4	

Figure 3.4 Word lists derived from documents 1, 2 and 3.

documents containing any three terms, down to documents containing only one of the terms. The boolean query gives a sharper cut of the documents, while the nearness query gives a more diffuse view.

a 4 D1	design D2	music 2 D3	significant D1
a 4 D2	developing D1	needs 2 D1	significant D2
a 3 D3	developing D2	needs D2	slots D2
acquired D2	engineer D2	of 3 D1	specifications 2 D1
and 2 D2	existing D1	of 3 D2	specifications D2
and D3	existing D2	of D3	system D2
archive D2	expressed D1	on D3	the 4 D1
as D1	expressed D2	organization's D1	the 6 D2
as D2	for D1	organization's D2	the 2 D3
assembled D2	for D2	our D3	their D2
be 2 D1	for 2 D3	pleasing D3	theme D3
be 2 D2	frames D2	portion D1	then D1
client D1	ideally D1	portion D2	then D2
client D2	ideally D2	program D3	which D1
compiled D1	identify D1	programmer D1	which D2
compiles D3	identify D2	programmer D3	which D3
consults D3	information D1	prototype D1	will 4 D1
contribute D1	into D1	prototype D2	will 4 D2
contribute D2	into D2	records D1	will D3
data 2 D1	knowledge 4 D2	records D3	
database D1	library D1	relationships D2	
database D3	library D3	rule-based D2	
design D1	make D3	selection D3	

Figure 3.5 Merged word lists from Figure 3.3.

acquired D2	express D1 D2	program D1 2 D3
archive D2	frames D2	prototype D1 D2
assembled D2	ideal D1 D2	record D1 D3
client D1 D2	identify D1 D2	relationships D2
compil D1 D3	information D1	rule-based D2
consults D3	knowledge 4 D2	selection D3
contribute D1 D2	library D1 D3	significant D1 D2
data 2 D1	make D3	slots D2
database D1 D3	music 2 D3	specificat 2 D1 D2
design D1 D2	needs 2 D1 D2	system D2
develop D1 D2	organization D1 D2	theme D3
engineer D2	pleasing D3	
exist D1 D2	portion D1 D2	

Figure 3.6 The index.

The differences therefore come mainly in how they are used in conducting a search. Using a boolean language, one typically starts with one or two terms. The ISAR responds with the size of the response set. If this is too large, the user refines the query by adding another term using *and* or *and not*. If it is too small, the user can broaden the query by adding another term using *or*. When the result set is small enough, the user then requests that the set be displayed. Many systems have a facility to store and name result sets so that they can be combined in this way. It would be poor practice for a user to begin searching with Query 1 in a boolean ISAR – that query would be arrived at after a series of refinement steps.

On the other hand, Query 3 would be a good starting point if a nearness-based ISAR is used. If the nearest document contains only three of the terms, that document would be first in the response, so would be seen. Generally, the more terms in the query, the better. Nearness ISARs often allow a whole document to be included in the query. The searching strategy is to start the query using several terms as in Query 3, look through the responses to find a relevant document, then including that document in the query for a further search. This technique of query refinement is called *relevance feedback*.

Following this example through, a nearness-based ISAR would respond to Query 3 with Documents 1 and 3, since they both have 100% nearness to Query 3 (they have all the terms in Query 3, and the nearness is normalized by the number of words in the smaller document). Assume that Document 1 is judged by the user as relevant, and the ISAR is asked to retrieve documents near to Document 1.

Document 1 has 21 words, Document 2 has 24, and Document 3 has 11. Document 1 has 14 words in common with Document 2 and 5 words in common with document 3. Documents 2 and 3 have no non-stop words in common. Therefore Document 1 has a nearness from Document 2 of $14/21 = 67\%$, and a nearness from Document 3 of $5/11 = 45\%$. Document 2 therefore becomes nearer to the query than Document 3. Document 2 is relevant, so that the recall of the query is improved.

Nearness has in this case retrieved a relevant document that contains none of the original key terms – this requires some explanation. What has happened is that both Documents 1 and 2 contain the terms {client, contribute, design, develop, exist, express, ideal, identify, needs, organization, portion, prototype, significant, specificat}, which are characteristic of design and development activity.

Note that the problem of the relationship between words and meaning discussed above has not been solved. Since the words in common are generic to design and development activity, the article could easily have been about designing a manufacturing plant rather than a computer system, so that a document with nearness 67% might have been quite irrelevant. Experience is that the percentage nearness figure given by many search engines has very little relationship with relevance to the user's information need.

This phenomenon can make nearness-based query languages difficult to use. It sometimes happens that many of the terms of the query occur in several contexts, for example almost any query concerning the Republic of China (Taiwan) is for documents for which there are similar documents concerning the People's Republic of China (government, economy, tourism, etc.). A query for "information management" (the care and feeding of information as an asset to an organization) would be swamped by responses about management information systems and other fields. In these situations, many irrelevant documents share the terms and are returned with a high nearness figure. Nearness languages work much better in collections of documents on a similar topic, where the meaning of terms is more uniform. It is ironic that their greatest use is in the most heterogeneous collections – Web search engines.

Nearness-based query languages are therefore somewhat different in performance from boolean query languages. The main advantage of nearness languages is their simplicity, while the main advantage of boolean languages is the facility they provide for the user to control what appears in the result set. In fact, most of the search engines permit hybrid queries that combine features from both languages.

A user's real need generally requires multiple queries

The foregoing discussion has implicitly assumed that a user's information need can be expressed in a single query, and the need can be satisfied from its response. In practice, this is rarely the case. Suppose one's information need is to support a paper outlining the possible impact of very low cost, high bandwidth communication technology on the accessibility of medical services in remote communities. If a paper on this topic already existed, it would not need to be written, so a query looking for that specific topic would not be expected to produce any useful material. The researcher would look for different aspects of the topic with different queries: one to identify trends in technology, one to identify trends in telecommunications company tariffs, and probably several to identify different information needs derived from particular medical problems of particular remote communities.

Alternatively, one might wish to find out the state of the art in Web-based search engines. A query on a bibliographic database might identify a few items of interest, some of which could be quite old – say the Harvest project at the University of Colorado. The user might then look for other work by the key authors, other papers referring to the project, and also find the Web pages of the Department having done the work and the agency supporting it, to look for other related material.

Furthermore, a particular document may be relevant to an information need not because it answers a predefined question, but because it reveals unsuspected aspects of a topic that the user employs as the basis of further queries. A search for information on making concept maps for design of a Web site (see Chapter 5) might return a Web site containing an atlas of cartographic representations of a wide variety of aspects of cyberspace, which might lead to a search for material on various cartographic principles. All but the simplest query topics in fact generally evolve.

This sort of higher-level search behaviour is sometimes called *berry-picking*. The user spends considerable time with a number of information sources making many queries of different types, from time to time recording some relevant material in a local workspace (the berry basket). An information space designed to support berry-picking would allow many different ways to navigate through the space using, for example, subject, author, footnotes, citations, journal run (the series of articles in a particular journal for a period of time), and area scanning (looking at similarly classified books in the library). Berry-picking can be done using a Web browser interface to a number of bibliographic databases and search engines, with the relevant items cut from the responses and pasted into a wordprocessing document or a database on the user's personal computer.

How do we know when to stop?

We have noted that while it is easy to estimate precision, it is very difficult to estimate recall. How do we tell how much of the relevant material we have retrieved?

There are several factors in this process: We need to know something about:

- the characteristics of the information we are seeking;
- the nature of our information need;
- the characteristics of the information source we are searching;
- the characteristics of search results.

Sometimes we know quite a lot about the information we are seeking. If we are looking for Microsoft's Web site, someone's telephone number, or the screen times of a particular film at a particular theatre, we know that there is only one relevant object, and know how to recognize it. If our information source returns that object, we know our recall is 100%, and need look no further.

Sometimes our information need is known to be limited. We might be looking for a photograph of a rhinoceros, a book on category theory, or for a Tibetan restaurant. We aren't looking for all such objects, or the best. We will find a few, then choose one and go on about our business, so we don't need a very high recall.

Other times we know the reputation of our information source. If we are looking for research papers on the side effects of Prozac, we know that Medline (operated by the National Library of Medicine in the USA) has every reputable scientific paper in fields related to medicine. We also know that the scientific literature in the medical and pharmacological fields use a highly standardized vocabulary. Once we figure out how Prozac is described, we are pretty sure we have all papers on the topic. Other sources are complete but not so well standardized. Virtually every significant event in a city is reported by at least one of its newspapers, so that we can be sure that a search of its archives can find reports about a particular event. The problem is how to describe the event in the query – we may have to make several different searches using different query terms (see the discussion of the principles of variety and uncertainty in the next chapter).

Finally, when we are searching for an indefinite body of information in possibly incomplete sources, we must take the berry-picking approach described above. In this case we must make many searches using different query terms and different sources. We might find many of the same objects coming up in different searches – that indicates a high degree of recall. We may find authoritative survey articles or results of extensive studies carried out by government departments which give us a good idea of the structure of the results we are seeking. If doing research, we may look for material prepared by well-known workers in the field – if our searches in information science turn up a number of works of which Gerard Salton is among the authors, then we can be confident of a reasonably high recall.

Obtaining adequate recall requires some judgment. If we are doing a search on behalf of someone else, we may have to argue for our judgment of the adequacy of recall as well as present the results of our research.

Key concepts

The proportion of retrieved documents matching the information need is called **precision**. The proportion of documents which would match the information need which are retrieved is called **recall**. Satisfaction of a user's information need often requires many queries, with several retrieving relevant items. This strategy is called **berry-picking**.

Further reading

 This chapter has concentrated on ISAR systems from the point of view of their users, and has not discussed the technology with which they are constructed. There are several excellent books concentrating on the technology, including Korfhage (1997), Salton (1989), and Van Rijsbergen (1979). Notice that some of these books are decades old. That these books are still relevant reflects the fact that the algorithms and processes used to support information retrieval have not improved much since the first systems were implemented in the 1960s, which in turn is a reflection of the difficulty of building systems that can operate in terms of meaning rather than words. The huge changes evident have come rather from increased power and decreased cost of the underlying computing, storage and communications technology.

However, a large number of advanced techniques are reported in the literature, many of them in the references cited. These are sometimes employed, but generally have results not markedly better than the basic approaches discussed in the chapter.

The concept of berry-picking was introduced by Bates (1989).

Major formative exercise

Purpose: gain and reflect on concrete experience in information seeking, which will help with understanding of the key concepts in this and following chapters.

Pretend that you are going to write a 3000 word essay on some topic, which is sufficiently narrow to discuss at most 10 items of information. A topic such as "information technology" is too broad – rather something like "use of mobile computers to assist recording of field data collected in ecosystem monitoring".

The items of information should be a mixture of research papers and World Wide Web pages. Search for information for this pretended paper using several sources, including at least:

- a bibliographic source which uses a boolean query language;
- a search engine on the Web such as Alta Vista which uses a document distance (nearness) based query.

a. Make a detailed record of your experiences, including:
  - the queries you made (text of the query, system query made on);
  - description of the results (this can be summarized);
  - the proportion of the results relevant to your topic (precision) at each stage;
  - the query processing method used by the source (boolean, document distance);
  - special features of the source used (saved sets, relevance feedback, etc.).
b. Present the topic for your pretend paper in a sentence or two. Comment on how the topic evolved during your search.
c. For two items retrieved which are relevant to your topic, comment on how you could tell that the item was relevant, with an emphasis on what information outside the item was needed to make the decision. Show the full text of both items if not too large (an abstract, say), or make an abstract of them.

d. For two items retrieved which are *not* relevant to your topic, comment on how you could tell that the item was not relevant, with an emphasis on what information outside the item was needed to make the decision. Show the full text of both items if not too large (an abstract, say), or make an abstract of them.

e. How confident are you that you have all the information relevant to your (remember quite narrow) topic that is contained in the sources you used (high recall)? Justify your opinion using evidence from your particular experience. Comment individually on both searches.

f. Make an assessment of how effective the various elements of the various query languages you used were in your search. Include the boolean connectives and, or, not, and the document nearness approach.

Your report should give general conclusions based on your specific experiments.

Tutorial exercises

1. Consider the following collection of publication titles and authors, selected from a library catalogue with the query *subject term = power and year = 1993.*

No.	Title	Author
S 0001	Aboriginal health and history power and prej	Hunter Ernest
0002	Advanced emission controls for power plants	
0003	Advanced coal-based power generation technic	Bharucha Noshir
0013	The application of expert systems in the powe	
0014	Asia-Pacific Economic Co-operation Working Gr	
0018	Coal solid foundation for the world's electr	
0023	Constitutional law in the Middle East the em	Mallat Chibli
0024	Desulphurisation 3	
0030	Electric power technologies environmental ch	
0038	Energy taxes and greenhouse gas emissions in	McDougall R.A.
0039	Energy policies of Romania survey	
0040	Environmental aspects of nuclear power paper	
0044	Flooding Job's garden	
0047	Gas engines for co-generation papers present	
D 0050	High-power Gas FET amplifiers	
S 0065	Iron, gender, and power rituals of transform	Herbert, Eugenia W.
S 0072	Judicial power and the charter Canada and th	Manfredi C
S 0074	Last rights death control and the elderly in	Logue Barbara J.
0075	Law, liberty, and justice the legal foundati	Allan T.R.S.
0076	The legislative process in the European Commu	Raworth Philip
0078	Maritime change issues for Asia	
0081	Museum of contemporary art vision & context	Murphy Bernice
0084	Nuclear power plant safety standards towards	
0085	Off-site nuclear emergency exercises proceed	
0087	A peaceful ocean? maritime security in the P	
0099	The power of one	
D 0100	Power electronics circuits, devices, and app	Rashid Muhammad
D 0101	Power electronics semiconductor switches	Ramshaw R.S.

D 0102 Power MOSFET design   Taylor B.E.
    0113 Purchasing power parities and real expenditur
    0119 Report on a review of police powers in Queens   Queensland
    0120 Report on review of independence of the Attor   Queensland
    0121 Report on review of independence of the Attor   Queensland
    0125 Rural and remote area power supplies for Aust
    0126 A short course on design of transmission line   Short Course on Desi
    0130 Spectrum estimation and system identification   Pillai S. Unnikrish
S 0131 Spectacular politics theatrical power and ma   Backscheider Paula
S 0134 Up against Foucault explorations of some ten
    0135 Visitor centres at nuclear facility sites pr
    0137 Work management to reduce occupational doses
    0138 World energy outlook to the year 2010

a. Consider the two subsets of entries labelled "D" (power electronic devices) and "S" (socio-political). For each entry, discuss how you know which category it belongs to, with emphasis on the evidence drawn from the title only and on what further information you used to make the decision.

b. Consider the four records labelled "D" (power electronic devices). Attempt to limit the search with at most two additional key words on the title to get as close as possible to those records (you may lose at most one). What is the precision of your resulting query, and the recall with respect to the records displayed?

c. Repeat the exercise on the six records labelled "S" (socio-political). You may lose at most 2 of the records.

2. Pick a small number of titles from the list in Question 1, which appear to be a coherent collection of information relevant to some specific problem. What queries would you need to retrieve them? What other sorts of information would you need and how would you find it? (The berry-picking approach.)

3. Consider the following collection of publication titles and authors, selected from the list in Question 1, as being directly related to electric power generation. Regard this list as the entire universe for the purpose of this question.

No.	Title	Author
0002	Advanced emission controls for power plants	
0003	Advanced coal-based power generation technic	Bharucha Noshir
0018	Coal solid foundation for the world's electr	
0030	Electric power technologies environmental ch	
0040	Environmental aspects of nuclear power paper	
0047	Gas engines for co-generation papers present	
0084	Nuclear power plant safety standards towards	
0085	Off-site nuclear emergency exercises proceed	
0135	Visitor centres at nuclear facility sites pr	

a. Show the full list index for the terms: power, nuclear, safety.

b. Show the intermediate and final index lists computed during the course of processing the query: *power* and *nuclear* and *safety*.

c. What terms in the sample set of titles might be considered as *stop words* (a stop word is a common or function word that is unlikely to contribute to any query, and is therefore excluded from the index).

d. Ignoring the stop words, show the document nearness of each entry from entry 0084, using the nearness measure: number of words in common/ number of words in entry 0084.

Open discussion question

 There have recently been introduced several advanced search engines said to be based on artificial intelligence and natural language processing techniques. Do any of these significantly improve precision and recall? If so, under what circumstances? How do they do it? Is there any a priori reason to expect unqualified success?

4 Describing information objects

This chapter explains how to describe documents so that they can be retrieved and the use of descriptors to gain an overview of a collection.

What is stored?

The purpose of this text is to facilitate understanding of the design of computer-based technology for access to and the management of collections of documents such as those in Table 1.1. The key observation from the point of view of this chapter is that most of the collections are of documents that are not stored in computer systems. Abstracts and e-mail messages are, but books, records/CDs, movies, software libraries, TV news footage, etc. are not. Other collections, such as museum collection catalogues, are actually not the documents of interest – the user generally wants objects in the museum collection, not the catalogue entry. The same is true of the abstracts and other information stored in bibliographic databases – the user does not want the abstract but the article abstracted.

In general, the primary documents, or perhaps better called *information objects*, in the collection are not suited to ISAR systems. They must be represented in the ISAR system by a specialized document called a catalogue entry or, more generally, a *surrogate*. The surrogate contains several kinds of information:

- a description of the content of the object constructed from a suitable vocabulary;
- identification of the primary document;
- possibly a physical description of the object.

This chapter focuses on a part of Figure 2.1, the relationship between the documents and the surrogates, as shown in Figure 4.1. The problem is the same as the problem of registration described in the Costigan Royal Commission case study in Chapter 2.

How is an information object identified?

A surrogate represents an information object, but is distinct from the object. The two must therefore be tied together in some way. The surrogate must contain an

29

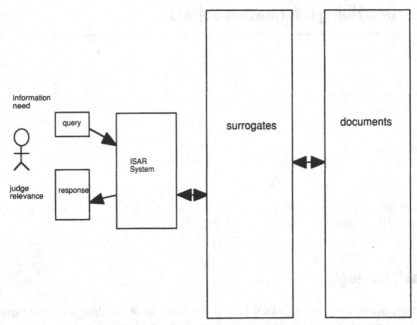

Figure 4.1 ISAR system relationship to collection of documents.

identifier for the object. This may seem obvious, but the problem is not so simple as one might think.

Consider, for example, a library catalogue entry. A typical catalogue entry will contain three different identifiers of the book for which it is a surrogate:

- author, title and publisher;
- library call number;
- International Standard Book Number (ISBN).

Each of these identifiers has a different purpose, and is designed to solve a different problem. Consider what the user wants an identifier for. One possibility is to go to the shelves and look at the library's copy of the book. For this purpose, the user will employ the library call number.

Alternatively, the user may want to recommend the book to a colleague located elsewhere in the world. In this case, the library call number is not adequate. Should the colleague wish to consult a copy of the book held in a local library, the call number will not generally be the same. If this book is in the colleague's personal library, there is no call number at all. For this purpose, the author, title and publisher form the appropriate identifier. Published citations to a work fall into this category, and make use of the author, title and publisher identifiers.

Finally, one might want to purchase the book. Here, the ISBN is a good identifier, since it distinguishes editions and different forms of binding. Booksellers, distributors and publishers use the ISBN to identify the book in their inventories.

In fact, the user will often identify the work by author and title in the query seeking the book's catalogue entry, even if their intent is to find it on the library shelves or to purchase it. This identifier can be considered as the primary public identifier, while the other two are essentially private identifiers within the library and publishing systems respectively.

The library catalogue example shows what we have to think about when we design identifiers. First, how does a user identify the object in seeking the surrogate?

Where the object is a copy of a published work, such as a book, CD, video, or piece of music, there is generally a public identifier similar to author, title, publisher. Librarians have gone to considerable trouble to develop means of identifying such items. Other published works, such as newspaper articles, do not have standardized public identifiers. They are found by searching the descriptors in the surrogate and linked to the published documents by essentially private identifiers such as paper, edition, date, section, page and headline. Arias in operas are identified by opera name, composer and the first few words, as are scenes in classic plays (e.g. Shakespeare's *Hamlet* "to be or not to be").

Collections where the objects are primary sources (Web pages, memos, images, TV news footage, etc.) often do not have public identifiers, and must be called for by terms from the content description. Private identifiers depend on the document and how it is stored – a Universal Resource Locator (URL) for a Web page, a file number for a photograph, scene and shot number for TV footage.

Documents from public enquiries, problem reports and so on, are the primary source, but have public identifiers – a submission to an enquiry is identified by submitter and date, for example.

In short, design of identifiers for an object in a surrogate requires taking into account how the user will look for the surrogate, and who will use the identifier and how.

Physical description of document

The surrogate will often contain some physical or quasi-physical information about the document. Surrogates for books often include the size, number of pages, whether indexed, etc. A piece of music will give its medium or file format and playing time. A newspaper article surrogate may contain its size in column centimetres or words.

Content description

Identifiers are essential in a surrogate, and physical descriptions are often useful, but the primary purpose of the surrogate is to be the representation of the document in the ISAR system. Therefore the primary design problem for a surrogate is how to describe the contents of its document.

Even though documents vary enormously in form and many kinds of documents are entirely non-linguistic, descriptions of documents are almost always given in text. Technological reasons are often given for this – we have long experience with text descriptors and have well -established techniques for fast searching of large sets of textual descriptors, whereas non-textual searches are much less well understood and are computationally much more expensive.

It is true that there are applications with image-based query facilities. It is possible to search a collection of fingerprints using a scanned partial print from a crime scene. It is also possible to search a collection of mug shots given a police sketch of a person drawn with the help of a witness.

The reason for the prevalence of textual descriptors is much deeper than technological, however, and applications where image-based query facilities are in principle desirable are quite specialized.

Imagine a human librarian with an intimate knowledge of a document collection, for example a photographic archive. Suppose you are looking for photographs to use as inspiration for a drawing of the typical costume of pioneer women in Iowa in the 1880s to be used as an illustration for a historical novel. How would you ask the librarian for what you want?

You would probably ask using a question like "Please give me some photographs showing rural scenes from Iowa in the 1880s containing women engaged in everyday activities". Suppose you had a picture from a family album showing an Iowa pioneer woman performing an everyday activity in 1884, but wanted more pictures. If you simply scanned the image into the ISAR system and pressed a button "more like this", what would the system make of the photo?

The best a human could do would be to recognize that the photo was of Prudence Morgan milking a Guernsey cow, with her back to the camera. The human might know that Mrs Morgan lived in Iowa from 1881 to 1912. Let us suppose that the computer could accomplish this.

Which pictures are "like this"? How about:

- Mr Morgan milking a Jersey cow in profile (taken from in front of the cow);
- an unidentified woman milking a Holstein cow in 1954, in the last Iowa dairy to milk by hand;
- an unidentified farm worker attaching a Guernsey cow to a milking machine in Illinois in 1972;
- Guernsey cows in pasture in Nebraska, taken in 1901;
- Mr and Mrs Morgan's wedding photograph;
- Mrs Morgan's gravestone.

All of these are arguably in some sense "near" the sample photo, but none of them respond to the user's information need. In fact, if the user brought the sample photograph to our knowledgeable librarian, they would probably end up constructing the textual query above and not using the sample photo as such at all.

Image-based query systems like fingerprints and mug shots work because they are extremely specialized, and are used for very limited purposes – the images in the system are highly constrained, and the basis of similarity is well defined. Each search is assumed to be the same, looking for similarity in the same way, and with no ambiguity.

One could imagine a system which would extract faces from a collection of photographs and process them into standard forms, creating a mug shot file automatically. Or perhaps pictures of cars, or of cows.

However, suppose one were looking for car chase scenes in action movies similar to one planned, or for gymnastic routines similar to the gold medal winning women's apparatus performance in the Sydney Olympics in 2000, or for play patterns in a particular sport similar to a particular one from a given game. Even if the video streams were processed such that car chase scenes, or gymnastic routines, or particular play patterns were identified, the definition of nearness would be little easier than in the pioneer women's costume problem above.

In practice, each genre of activity has a vocabulary with which instances can be described or criticized. Think of the commentary at televised gymnastic or sporting events. Rather than relying on the machine's judgment as to what is similar to what, it would be much more productive to generate descriptions of each event using its critical vocabulary, and include that description in the event's surrogate. A query would be expressed in the same vocabulary, and the criteria of similarity expressed using that vocabulary. Both the descriptions and query are now textual so that text query mechanisms must be used.

One might want computerized support for generating the descriptions, and support for the user's formulating a query, but the same kind of problem arises in constructing descriptors from text-based documents. However the description is generated, it will generally be textual.

Choice of descriptors

The purpose of having descriptors in the surrogate is so that people whose information need would likely be satisfied by the document will find that document in the response given by the ISAR to their query. This requires guessing what terms the user will put in the query. To guess well requires knowing something about potential users and their likely information needs.

Sometimes we can know quite a lot about the users. Suppose the collection of documents is the archive of a daily newspaper. One set of users might be strategic managers in a large corporation. They want to be alerted to news events which have potential to affect the company, either because they relate to products, to sources of material, to the legal or regulatory environment, or to political/ economic instability in countries which are either major markets, major production sites, or major sources of materials. The information group would have access to the relevant information about the corporation, and also the skill to judge whether an item was a serious, immediate problem or simply something relevant. The surrogate for an article would contain the parts of the corporation affected, the type of effect and the seriousness of the effect, together with a precis of the article.

Now suppose the set of users is the secondary school students in a large school system with a well-developed curriculum. The surrogate for an article would contain the curriculum element to which it is related, mainly for the use of teachers planning lessons. If the collection is aimed at students, it might include project topics to which the article could contribute.

The Costigan Royal Commission case study of Chapter 2 is a further example of a situation where a great deal is known about the users and their information needs.

On the other hand, sometimes we know very little about the potential users of an information service. This is not to say that one can develop a successful information service with no thought given to who might use it, but that the users might be very diverse and little can be known a priori about their individual requirements. A newspaper archive maintained by the publisher could be such a case. Here, the material might be used by secondary school students, market researchers, strategic decision-makers, journalists, social researchers, genealogists, historians and many others.

Where the users are very diverse, the catalogue entries cannot be tailored to their requirements, so that the descriptors chosen must be generic to the material and to the methods by which it is produced. For newspaper articles, we can include terms from the text of the article – names of people and places, dates and terms which carry information about the topic. An article with the headline "Interactive TV 'in six months'", about trials being carried out in Australia for a form of interactive television involving broadcast video and telephone link back to the source, contains the names of two company executives, the names of several telephone companies and computer database suppliers, and the city of Sydney, Australia. Topic terms include "telco", "Enhanced TV", "modem", "Web service", "free-to-air telecasters", "digital broadcasting", "Internet" and so on.

A news article is generally news partly because it relates to large institutions, industries, events or trends. Which events and trends the article is related to are known at the time the article is published; indeed, these relationships are often part of the reason the article is considered news. It would make sense for terms designating these more contextual aspects to be included in the surrogate even if they do not occur in the article directly. For the example above, more general terms might be "datacasting", "television", "telecommunications", "electronic commerce", "multimedia", etc.

A photograph from a photo archive has nothing but contextual information in its surrogate: date and place taken, names of any people in it, name of the location and event, name of photographer, place published, identifiers of any prominent objects etc. Mr and Mrs Morgan's wedding photograph would have their names, "wedding", "Des Moines, Iowa, USA", "January 1881", "St Paul's Church", "Ebenezer Jones (photographer)".

Properties of descriptors

The particular descriptors chosen for a particular document depend on the document and the application, and so there is very little one can say in a general text like this. However, descriptors have properties which are independent of particular applications, and consideration of these properties is the added value of information science. There are two key properties: specificity and exhaustivity.

Specificity is a property of a particular descriptor with respect to a particular collection. It is a measure of how many documents in the collection contain that descriptor in their surrogate. There are many numerical measures of specificity, but they all have the property that as the number of documents containing that descriptor decreases, the specificity increases. The minimum specificity therefore occurs if a descriptor occurs in all documents, and the maximum if in only one. The opposite of specificity is generality – a descriptor is more general if it is less specific, and vice versa.

One possible measure of specificity of a descriptor is the ratio of the size of the collection to the number of documents containing that descriptor. So a descriptor occurring in 1% of documents would have a specificity of 100, 0.1% a specificity of 1000. That measure can lead to large numbers in large collections – a descriptor indexing only one document in a collection of 1 million would have a specificity of 1,000,000. A better measure might be logarithmic, so that a descriptor occurring in 1% of documents would have a specificity of 2, 0.1% a specificity of 3, and one in a million a specificity of 6.

Exhaustivity, on the other hand, is not a property of a single descriptor, but a property of a set of descriptors. A set of descriptors is exhaustive with respect to a collection of documents if every document's surrogate contains at least one of the descriptors in the set.

It is important that a set of descriptors be exhaustive. A user is going to query the surrogates in order to find documents, so if a document has no descriptors it will never be found, and might as well not be in the collection.

The two properties are related. An exhaustive collection of more specific descriptors will be larger than an exhaustive collection of less specific descriptors. If a collection of 10,000 documents has 100 equally specific descriptors, and each document has exactly one descriptor in its surrogate, then the average specificity is 2 (logarithmic measure). If we want to increase the average specificity to 3, then we need 1000 descriptors. A collection of 1 million documents with a set of descriptors occurring in 10 documents will each have an average specificity of 5, and 100,000 descriptors will be needed.

Engineering issues in surrogate design

Like any technological object, a set of descriptors comes at a cost and produces a benefit for some customers. There are several ways to approach building descriptors, each of which has a different cost and leads to a different system with different benefits to the customer. Benefits depend on the customer, so are mostly outside the scope of information science, but the different design approaches which determine cost can be examined.

A set of descriptors must be built, and then it must be updated as the population of documents evolves. It must also be used: by the service, who puts descriptors into surrogates for documents; and by the customer, who must successfully choose descriptors to find the document.

There are two main ways to obtain a set of descriptors: to devise a set by human effort, or to automatically generate the terms from the documents. The human-devised descriptors are generally inserted into the surrogates by human indexers, while the automatically derived terms can usually be automatically inserted into the surrogate of a new document. In either case if the new document is not adequately described by the existing set of descriptors, the set of descriptors must be updated, either by hand or automatically, according to the strategy by which the set was built.

Devising and updating a set of descriptors by human effort is far more expensive than generating and updating one by computer. Also, it is far more expensive to assign descriptors to a document by hand than automatically. The obvious question is to ask why anyone would consider any strategy other than full automation. To answer this question, we will first look at the process of automatic generation to see its capabilities and limitations. A program which generates a set of descriptors from a set of documents has as its input the full text of the documents and essentially nothing else. It is easy to write programs which identify the words occurring in each document, together with the number of times each word occurs. Figure 4.2 shows such a tabulation for Documents 1, 2 and 3 of Chapter 3, with the stop words removed. It is slightly harder to recognize phrases like "information retrieval", but quite feasible.

One approach is to simply accept these lists as the respective descriptors of the documents. The complete set of descriptors for the entire set of documents is simply the union of the individual lists, probably stemmed so that for example "compiled" (Document 1) and "compiles" (Document 3) are not distinguished. Is there any need to go further than this? Many systems in fact do not, but some systems recognize that some words are better descriptors than others.

D1	D2	D3
client.	acquired	compiles
compiled	archive	consults
contribute	assembled	database
data 2	client	library
database	contribute	make
design	design	music
developing	developing	pleasing
existing	engineer	program
expressed	existing	programmer
ideally	expressed	records
identify	frames	selection
information	ideally	theme
library	identify	
needs 2	knowledge 4	
organization	needs	
portion	organization's	
programmer	portion	
prototype	prototype	
records	relationships	
significant	rule-based	
specification 2	significant	
	slots	
	specifications	
	system	

Figure 4.2 Tabulation of words and frequencies in Documents 1, 2 and 3 of Chapter 3.

Descriptors are the terms used by the user to formulate a query, which the ISAR system will process to get a reasonable-sized result set which should have reasonable precision and recall. To facilitate recall, the descriptors should represent the documents' content well. To facilitate precision, the descriptors should distinguish among documents. These two requirements are to some extent contradictory.

For example, suppose the document collection were newspaper clippings referring to the University of Queensland. One would expect that nearly all the documents would include the words 'university' and 'Queensland', or the phrase 'University of Queensland' if the tabulation were more sophisticated. These terms are good descriptors of content, since all the documents are in fact something to do with the University of Queensland, but don't distinguish documents very well – a query using these terms would return nearly all documents, having great recall but very poor precision. These terms have very low specificity. Some systems exclude from the set of descriptors terms that occur in more than a nominated percentage of documents (which by the way automatically excludes stop words).

Some systems, especially those designed for document nearness-based query languages, weight some terms more than others on the grounds that they describe the content better. A simple way to do this is to weight terms in a document by the number of times they occur in that document. The term "knowledge" in Document 2 of Figure 4.2, for example, would have a much higher weighting than "specification". The term "specification" would have a higher weighting in Document 1 than in Document 2. There are many other, more elaborate, schemes for generating weightings.

The biggest weakness in automatically generated descriptors is that they are limited to the terms that actually occur in the documents. The "Interactive TV…" news article example discussed above is a good example of this weakness. The automatic approach would not be able to include the more general terms "television", "telecommunications" etc., nor the only slightly more general term "datacasting", since they don't occur in the text. The speech by Meagher referred to in Chapter 2 does not contain the term "Costigan Royal Commission". His audience would have known beforehand that Meagher was a central figure in the Commission – that would have been why he was invited to speak. All this context is lost in the document archive, so human indexers are often employed to add it.

The cost of human indexing is roughly proportional to the size of the index. The biggest index is at the highest level of specificity, which is the level of specificity of the automatically-generated index. Human indexing is used when it can produce surrogates sufficiently more useful to the customers of the ISAR system to justify the cost. Experience has shown that the effectiveness of highly specific human-generated indexes is about the same as that of automatically generated indexes. However, the more general indexes which the automatic systems are incapable of providing are very useful. Since they are less specific, they are also much smaller, so are much less costly. Hybrid systems are therefore in wide use – bibliographic systems use as descriptors both the full text of abstracts and also more general subject descriptors added by human indexers.

Other tools like synonym lists are also provided by humans. Of course, where the documents are not textual, the descriptors must be generated and applied manually. On the other hand, computer analysis is generally used to test sets of descriptors for exhaustivity.

Issues in the design of human-constructed subject descriptors are considered much more deeply in Chapter 12.

Bates's principles: Variety and uncertainty

The issues involved in designing surrogates can be further understood using two principles introduced into Information Science by Marcia Bates: variety and uncertainty. At bottom, the design of surrogates is governed by several factors.

- Most fundamentally, the size of the set of relevant documents. This is governed by the ability of a human to absorb and understand the documents, and then to synthesize them in some way, depending on the task at hand. A student writing a term paper needs a few tens of documents, while a PhD thesis needs a hundred or so. It is acceptable for either to miss a few documents, so long as they are not of central importance. (The recall need is high, but not 100%.) However, their information need is not an arbitrary selection of documents from a large set –

their topic must be specific enough that the set of relevant documents is of the right size.

- Next, the size of the set of documents the user is willing to scan to find relevant documents (minimum precision tolerated.) This is also governed by human ability, in this case to skim titles and document summaries. This determines the size of result set the ISAR will be expected to produce.
- Finally, the size of the collection.

The *principle of variety* is that the descriptors must be sufficiently specific to retrieve a result set of the appropriate size from the collection. This suggests that if the collection is small, very general descriptors are sufficient, while for very large collections even terms as specific as occur in the document may be insufficient, so that multiple descriptors are needed.

For example, consider several situations where identifying people is an issue.

- An auction. There are rarely more than a handful of bidders for any particular item. The auctioneers identify bidders by general descriptions like "the man to my left" or "the lady with the blue hat". Everyone knows who is meant, and the system works quite well.
- A tutorial or seminar class with 20 participants. Here, first names are generally used. There are sometimes two people with the same name, but which one is being referred to can generally be inferred from context: "Bob has just said..." refers to the Bob who has just spoken, as distinguished from the other Bob who hasn't said anything for 10 minutes.
- A large class with a few hundred students. Here, even surnames do not have sufficient variety to identify a person. The combination of first name and surname may even fail, in which case a middle name may be required to disambiguate.
- The student body at a large university. Here, among tens of thousands of people, a person must be identified by full name, date of birth and perhaps postal code of home address. (Date of birth is not very specific in this population because it can be expected to have a large proportion of members in a limited age range, so many people will share birthdays.)
- The population of the United States. Among several hundred million people, name, postal code and date of birth may not be sufficiently specific. Other identifiers such as mother's maiden name, place of birth, and occupation may be needed.

As the size of the population increases, the descriptors in the surrogate for a person must first increase in specificity, then multiply, so that a query can be constructed which identifies a single person, or at worst a small group which can be disambiguated in a further investigation.

Application of the principle of variety leads to more specific descriptors as the size of the collection increases. (We can think of a combination of descriptors as a single more specific descriptor.)

The *principle of uncertainty*, that more specific descriptors are unreliable, puts a limit on this process. The principle of uncertainty derives from the fact that in an ISAR system, attachment of a descriptor to a document must happen at least twice: once by an indexer, and once by a person for whom that document is relevant. In practice, most users want several documents, so that the same descriptor would ideally be attached to all of them; and most documents are relevant to many users over a period of time. The question is therefore how consistently a collection of people can perform the indexing task over a period of time.

The answer is "not very". Studies have shown that even professionally trained indexers differ in the descriptors they assign to the same document. It is certainly true that different documents often describe similar things in very different terms – Documents 1 and 2 of Chapter 3 are an excellent example. Even a single person may describe a single document differently over time.

Many people maintain files of articles they have found relevant to their research. These files may accumulate to a few thousand documents over a long time. It can be convenient to maintain an ISAR system to index this private collection, so whenever a document is added, the researcher creates a surrogate in the ISAR system and whenever some documents are needed the researcher queries the ISAR system.

However, a person's research changes over time. Suppose a journalist comes across the Meagher speech of Chapter 2 while working on an article on criminal investigation technology. Since the journalist's private collection of articles is fairly small, the principle of variety suggests that a fairly general descriptor is sufficient, so the descriptors "criminal investigation" and "technology" are used in its surrogate. Suppose some years later the journalist is working on an article on the applications of information retrieval software, and makes the query using the descriptors "information retrieval" and "application". The Meagher speech is relevant, but will be missed.

Reliability of indexing is greater for more general descriptors. The Meagher speech is more reliably described under say "law" and "computing" than the more specific descriptors in the previous paragraph.

The interaction of these two principles suggests that a good indexing strategy is to assign multiple descriptors to a document (cross-referencing), and more alternative descriptors as the descriptors get more specific. It also suggests that a user of the ISAR system is advised to make multiple queries using different descriptors, again with more alternatives as the descriptors get more specific. In fact, it is a good indicator of high recall if some of the same documents are returned by queries using different terms.

It is possible for the ISAR system to assist with communication among the various people describing documents by, for example, tabulating how often pairs of descriptors co-occur in documents. Some Web search engines do this. If, following the principle of uncertainty, the Meagher speech is described using all of the terms' descriptors – "criminal investigation", "technology", "information retrieval" and "application" – then when an article on, say, an investigation into possible criminal behaviour of the President of the United States is to be indexed and the indexer chooses "criminal investigation" as one descriptor, the ISAR system can suggest that "information retrieval" or "technology" or "application" be considered. Similarly for someone making a query using one of those descriptors.

This sort of approach is clearly more practicable for more general sets of descriptors. With highly specific sets, the human is likely to be overwhelmed by alternatives.

Push technologies

So far, we have implicitly treated the problem of describing a document in a surrogate as different from describing a desired document in a query. In fact, the two problems are very similar.

When one makes a query to an ISAR system, the response is a collection of documents selected from the collection in the system at that time. However, the information need may be continuing. Suppose someone wants to monitor a newspaper for stories about electronic commerce in book distribution. A query on the archive on 30 April 1999 will get the stories published up to that date, but what about later?

Bibliographic systems have long offered alerter services, where queries from subscribers are stored in what amounts to an ISAR system. As new documents are added to the archive ISAR, they are presented to the query ISAR as queries. The response to a particular document is the set of subscribers' queries which that document matches. In this way, whenever an article indexed by the terms "electronic commerce" and "book distribution" is added to the archive, the subscriber looking for stories about electronic commerce in book distribution will be notified. Perhaps the ISAR system will generate an e-mail containing the story.

This sort of service is increasingly becoming available from Web sites, where it tends to be called *push* technology (the information is pushed out by the source as opposed to being pulled in by a user making a particular query at a particular time). Push subscription is offered by news sites, by book retailers, and by employment agencies, in addition to bibliographic services. In contrast to push, the sort of queries described in Chapter 3 are sometimes referred to as *pull* technology.

Content analysis

Until now, we have concentrated on indexing documents so that they can be found individually by ISAR users making queries. Another use of ISAR technology is to create an overview of the content of a collection of documents. This is called *content analysis*.

Content analysis is performed by media analysis organizations interested in the range of topics covered in the press during certain periods, and how this content changes over time. A government health department might be interested in how much and what kinds of health information is published in a range of women's magazines, and how this has evolved over a period of years. This analysis might tell them how important health was perceived over time by these magazines' target markets, and what health issues were perceived as important.

It is also used by social scientists to analyse records of undirected interviews or observations of behaviour. A sports team may analyse videos of games, and be interested in how often certain types of player interaction occur as the season progresses.

All of these analyses work with surrogates for the underlying events. Each event is assigned a surrogate containing descriptors of the aspects of interest. Instead of a user making a query to identify particular events, however, the system produces totals of how often each event occurs, perhaps broken down by time. The absolute and relative frequency of occurrence of instances of each descriptor is the desired result. For example, the analysis of women's magazines might show that health items occurred only rarely in the 1940s and 1950s, then progressively more frequently up to the 1990s, where 25% of articles concerned health. More specifically, the analysis might show that before 1970 health issues mostly related to children, whereas topics such as AIDS and Alzheimer's disease had become more prominent in the later period.

Design of descriptors for content analysis poses somewhat different problems to that of design for information retrieval. In information retrieval, the critical factors are the size of the set of relevant documents and the size of the result set that a user is willing to tolerate. For content analysis, the critical factor is the size and complexity of the set of descriptors.

Consider a media analysis organization monitoring 500 daily newspapers in the United States and Canada. Each newspaper publishes perhaps 100 articles each day, so that in one year there are 15 million articles. By the principle of variety, an ISAR system would need hundreds of thousands of descriptors. Each descriptor would occur in hundreds of articles. A tabulation of the relative frequency of occurrence of hundreds of thousands of descriptors would, however, be incomprehensible, especially if it were to be analysed over time or by region. It would be hard to make sense of 1000 descriptors – a more comfortable number would be 100.

For these sorts of problems, very general descriptors are needed. For the women's magazine application, descriptors like "health", "craft", "recipes", "fiction" would give an overview. However, that example shows that more specific analysis is also wanted. To facilitate this, the most general descriptors are disaggregated into more specific ones; say, "health" is associated with "infants", "children", "mothers", "men", "women", "aged". Each of these would be disaggregated with further descriptors at the specificity of particular diseases or syndromes. The result is a hierarchically decomposed set of descriptors. The most specific descriptors could be the descriptors for the information retrieval application. Design of hierarchical sets of descriptors is discussed in some depth in Chapters 11 and 12.

The user can analyse the collection at a variety of levels of specificity, with the more specific breakdowns restricted to certain more general descriptors. In this way, one can first see how prominent health is compared to other general topics, then within health how the focus of concern within the family has evolved, then within a particular category, what specific issues are addressed.

This sort of analysis is becoming increasingly popular under the rubric "data warehousing". We return to the topic in Chapter 9.

Key concepts

A catalogue **description** of a document contains a selection from a set of descriptors. The whole set must be **exhaustive** with respect to a collection – all documents in the collection must be able to be described by at least one of the set. Each descriptor is associated with a proportion of documents in the collection – the fewer documents the higher the descriptor's **specificity**. The **principle of variety** is that there must be enough different ways to describe documents to make the distinctions needed. The **principle of uncertainty** is that different people may describe a given document in different ways. **Pull** is when a user makes a query on a fixed set of documents. **Push** is when a user establishes a filter query on a future stream of documents. **Content analysis** describes a collection of documents by how often each of a set of descriptors occurs in the collection.

Further reading

More technical treatment of automatic generation of descriptors can be found in Korfhage (1997), Salton (1989), and Van Rijsbergen (1979). The principles of variety and uncertainty were introduced into Information Science by Bates (1986). More detail on content analysis may be found in, for example, Weber (1990). Content analysis is the basis for a widely read series of studies of large-scale social trends by John Naisbitt and others with titles including the word "Megatrends".

Formative exercise

Investigate a variety of information sources, including some non-textual ones.

a. How are the documents identified?
b. How are the documents described?
c. Is the description aimed at particular audiences?
d. How specific are the descriptors?
e. Are they arranged hierarchically?
f. How are they generated?
g. Are the descriptors effective (can you use them successfully to find what you want in the collection)?
h. Would it make sense for the source to offer a push option? Does it?
i. Would it make sense for the source to offer an aggregate view of its contents? Does it?

Tutorial exercises

1. Consider the following collections of information objects:

- a newspaper;
- the mailing list for a course (things posted to it);
- a public enquiry into higher education in a country;
- television news items;
- spikes (Appendix L).

For each, discuss the following:

a. How would you identify the information object?
b. Who wants to find information from the collection, and what would a typical query be?
c. What would you use to represent the content of the object in the catalogue record (index, surrogate)?
d. How does the size of the collection affect the descriptors you use?

2. Discuss whether and how push technology and overall content analysis is relevant to each domain.

Open discussion question

Consider various collections of information, for example:

- the editorial part of the computer section of a major national daily paper;
  the job ads in the computer section of a major national daily paper;
  the income tax Act of some country;
- other collections you know something about.

Would you index them or just use free-text searches? If you would index, would you use a manual or computer-generated index? How would you decide?

5 Browsing: Hypertext

This chapter discusses the concept of browsing as distinct from searching, implemented in hypertext as for example World Wide Web sites or as CD-ROMs. It introduces the concept of information spaces.

Information spaces

In previous chapters we have been looking at the problem of finding relevant information in a collection of documents. We have seen that we can consider the documents to be related to each other in various ways – not only the concept of document nearness based on shared terms, but in the berry-picking section of Chapter 3 a variety of specific relationships among documents. That section introduced the concept of moving from document to document via these relationships.

The metaphor of moving suggests a larger metaphor of a space within which one moves. We are familiar with all sorts of spaces.

- Geographical space – moving around on the surface of the earth. We have orientation – north, south, east, west – and distance, supported by maps and various forms of transport.
- Urban space – moving around in a city. Besides the orientation and distance available because a city is a geographical space, we have neighbourhoods, roads, the public transport network, addresses, shopping, commercial, residential, leisure and industrial districts; aided by specialized maps, directories and signs.
- Architectural space – moving around in a large building. We have floors, corridors and rooms with addresses; public spaces such as lobbies, cafeterias and meeting rooms, supported by elevators, directories and signs.

In this chapter we want to consider a collection of documents as a space – an *information space*. We will begin with an example: a textbook such as this. For this exercise, the documents will be the lowest-level subsections. Each subsection is about a specific topic. Being at a point in the space means viewing the subsection, either as a page in a book or perhaps on a screen.

Points (subsections) near where one is include:

- next and previous subsection;
- subsections cross-referenced in the present subsection;
- figures or tables cross-referenced;
- citations of references;
- footnotes;
- subsections containing key terms contained in the present subsection;
- definitions of key terms.

So we can navigate around this space.

In the geographical spaces above, navigation is assisted by large-scale information structures and long-distance navigation aids. The comparable auxiliary information structures in a book include:

- physical layout – it is easy to see where you are in the book's linear sequence;
- table of contents – use it to find a general topic in context;
- index – use it to find a specific topic out of context;
- bibliography – from the text to its sources;
- glossary – definitions of key terms;
- some books have marginal notes for cross-reference or as mini-abstracts;
- some books have navigation paths showing which chapters depend on which.

This is how it is possible to see a book as an information space. Using these structures one can find one's way around the material and also find particular information. These structures also permit a degree of overview of the book-space. The table of contents encapsulates what the book is about. The index or glossary gives a good idea of the key topics. The bibliography tells us the intellectual traditions the book is derived from – are the citations mostly from information science or database systems? American or European?

Hypertext implementation of information space

A hypertext implementation of an information space stores the text in a computer system and allows the user to interact with it via computer-generated displays and an input device (generally a screen and a mouse). So a hypertext system consists of a collection of *pages* connected by *links*. The pages are analogous to places, and the links to paths between the places.

The information space can be the same as a book – what has changed is not the content, but the medium on which the content is published.

In some respects, the hypertext medium provides similar facilities to a book:

- pages are displayed rather than turned to and looked at on paper, but we still see a page;
- one can navigate from page to page, from section to section;
- one can return to the table of contents, then pass to another section;
- one can search the index, then pass to a section containing some keyword;
- one can refer to the glossary;
- one can pass to cross-referenced material.

However, in other respects, the medium used for hypertext works differently from a book.

- Access is much faster, making digressions from a linear reading much easier.
- Pages are generally much smaller and scrolling is more difficult, so information must be organized into smaller fragments.
- Smaller fragments and easier digressions mean that it is more difficult to see the whole, and easier to lose context. When the user loses context, this is analogous to being lost, since the user doesn't know what links to take. This phenomenon has been called *lost in hyperspace*.

The information space is held together for the user by the auxiliary information structures. Because of the danger of the user becoming lost in hyperspace, a hypertext needs more of these auxiliary structures than a book. Some extra structures possible include:

- deeper tables of contents, which can be dynamically expanded and contracted in depth;
- guided tours, which are recommended paths through the pages for particular purposes;
- bookmarks, so that a user can keep track of interesting or familiar places in the space;
- trace of navigation, so a user can keep track of how they got to where they are, and be able to re-trace a path;
- facility to search for pages containing particular terms – an extension of the index – using the tools discussed in Chapter 3.

Most of these should be familiar to the reader.

The medium is still new and the subject of considerable research and development, so that additional facilities should be expected in the future. Some other more experimental features are discussed below, and also in Chapters 9 and 13.

Building a hypertext

We have assumed that the reader is familiar with and has used maps, books and hypertext systems, so the exposition has been in the way of reminding – pointing out particular aspects of experience – and making comparisons. We turn now to the less familiar insides of hypertext systems – first, how to design them.

At the lowest syntactic level, a hypertext system is simply a collection of documents with links between them. At this level, the design problem is what to link to what. At almost this level, it makes sense to divide the documents into primary documents, those that the user wants to see; versus auxiliary documents, whose role is to assist the user to find the primary documents. These auxiliary documents are the auxiliary data structures referred to in the preceding sections.

What needs to be linked to what depends very strongly on the content of the hypertext, the primary documents, and on the primary audience the hypertext is designed to serve, so that it is difficult to give specific advice in a general text. Some general advice is possible, however.

The primary documents must be suited to hypertext display. That is, they must be small enough to be read without much scrolling. Moreover, they are generally best organised like a news story, with the main points visible without scrolling, allowing some users to skim the surface. Those users who want more depth can either scroll, as one would in reading a news story possibly following its continuation to an inside page, or follow links.

If the primary document is textual, some rewriting is often called for. For example, the section of this text you are reading now has a structure intended to be suited to someone reading a printed book. The first paragraph in this section is a link from the earlier sections to this section, where the material is presented differently. The second paragraph is a definition of terms, and the third is an apology. None of these contribute much to the 'news value' of the section, so that if this section were implemented in hypertext rather than a book, the very valuable space they occupy at the top of the page would be wasted.

A better beginning for a hypertext representation might be a mini-abstract such as:

> Hypertext design involves design of individual pages, auxiliary structures, and the links among pages, within the structures, and between the two. Some links between pages are structural links between different types of pages, while other links are semantic (based on shared content). Computer support is possible to help with creation of both structural and semantic links, but the process cannot be fully automated in our present state of understanding.

Paragraphs one and three would be probably omitted, since they make sense only if the reader has just finished the earlier sections, so the next paragraph might be paragraph 2, "At the lowest syntactic level …"

Some users might then scroll through the section. Others, however, might want more information on one aspect only, or in some order different from the sequence in the document. To make this possible, it would be a good idea to place anchors in the text and links in the first paragraph, which might now look like:

> Hypertext design involves design of <u>individual pages</u>, <u>auxiliary structures</u>, and the links <u>among pages</u>, <u>within the structures</u>, and <u>between the two</u>. Some links between pages are <u>structural links</u> between different types of pages, while other links are <u>semantic</u> (based on shared content). <u>Computer support</u> is possible to help with creation of both structural and semantic links, but the process cannot be fully automated in our present state of understanding.

If the primary documents are detailed statistical breakdowns, then the top might best be some summary totals, allowing the user to drill down to the detail. For example, Table 5.1 is derived from a detailed breakdown of trends in demand for computing skills from analysis of the job ads section of a major newspaper.

Table 5.1 Sales by region by quarter by product group, in million AU$.

	Average 1988	Average 1993–94	Average 1995–96	1997–98	1997	1998
Total Tech	86	73	93	91	90	91
Pre/Post Support	4	3	2	1	1	1
Sales/Mgt/PMgt	10	9	5	8	9	8
Grand Total	100	100	100	100	100	100

The three underlined headings in the left column of Table 5.1 link to more detailed breakdowns of which that row is the total.

Finally, a large graphic might be displayed as a thumbnail at a low resolution so that the user can get an overview of it, then link to more detail.

Primary documents often have a structure, so that links are needed to represent it. For example, in a technical reference the primary documents are the lowest level subsections, where the text often connects to figures, tables, examples, citations, definitions of terms, other sections and perhaps tutorial-type exercises.

Auxiliary documents are often highly structured, and this structure must be represented by links and other dynamic features. A hierarchical table of contents or site map might to be able to expand and contract, but certainly needs to have links to and from the primary documents. Indexes or glossaries of key terms often have an alphabetic table of contents, so the user can click on 'F' to view the keywords starting with 'F'. There must be links from the index to the primary documents, from the primary documents to the glossary, and from the glossary to the index.

The possibility of automatic creation of a hypertext

Building these links is a major job. A hypertext project often starts with the primary documents, and often with auxiliary documents derived from a print version of the material. What prospect is there for automatic creation of the hyperdocument? There are two aspects to this problem: the structural links and the semantic links.

In an already existing document, the structural links are represented in some way. In some cases (more and more), the structural links are represented using a comprehensive markup language like SGML or XML (see Chapter 7). In this case, the hypertext implementation can be generated by a special rendering process and can be fully automatic.

Otherwise, the structure is represented by various textual means:

- sections may be numbered 1, 2.2, 3.5.2, etc. or indicated by patterns of font, size, style and spacing;
- citations may be indicated by [xx] or (Surname, Year);
- index entries indicate a page number, so a program can match the keyword against the text, taking account of possible variants by stemming;
- links to glossaries can be identified by cross-referencing the keywords in the glossary to the keywords in the index, then putting the link to the glossary at the occurrences of the keyword;
- first occurrence of keywords may be indicated by italics.

However, these programs are unlikely to be completely reliable. After all, this problem is an instance of the information retrieval problem, and we know that it is practically impossible to attain 100% precision and 100% recall.

- Unless the document is very thoroughly edited and proofread, some structural elements may not be indicated by the standard means, leading to a failure of recall.
- Some stylistic features may be used for more than one purpose, leading to a failure of precision.
- Patterns of content may occur by accident. Numbers in tables may look like section numbers. A year may occur in a parenthetical remark.

- Searching for keywords is particularly a variant of the information retrieval problem.

If structural links can be difficult to automate, semantic links are more difficult. The present section contains links to at least two other parts of the book – explicitly to Chapter 7 in the mention of XML and implicitly to Chapter 3 in the mention of the 'information retrieval problem'. In both cases, one would like a link to a specific part of those chapters where the specific topic is discussed: in the first case the use of XML markup for hypertext documents and in the second the difficulties of achieving 100% precision and recall. To do this fully automatically with nothing other than the text you see would be a complete solution to the information retrieval problem, and the world would be a different place. Such an achievement is far beyond present technology.

There is plenty of scope for partial automation, however, for use of computerized tools to assist humans in the task of building a hypertext. We can do all the tasks proposed to fully automate the process, but instead of directly creating the hypertext we can present the results to a human, who will edit the results into hypertext. An ISAR system is a useful tool in turning a collection of documents into a hypertext.

Further text processing technology can also be useful. For example, in creating an index or in cross-checking documents written by different people for consistency of terminology, one needs lists of key terms with the location of their occurrence. It is easy to compute a list of distinct words in a body of text together with the number of times they occur, as in Figure 4.2. This list will contain the words used in key terms together with a large majority of words not in key terms. An editor familiar with the subject of the document can easily delete the 90% of unwanted words using a simple wordprocessor.

It is then easy to process the text again, extracting the occurrences of the remaining words together with fragments of the sentences in which they occur, together with the location in the text of the occurrence. This list will contain all the key terms, together with a large majority of phrases not containing key terms, or at least not significant occurrences of key terms. Again, using a wordprocessor an editor can delete the 90% of unwanted occurrences, then edit the remaining phrases into key words by removing unnecessary words and standardizing the presentation.

For example, if such a process were run on this text, the word occurrence/ frequency tabulation would contain the word "information", "retrieval", "problem", and "problems", among many others. These words would be retained in the first editing pass. Figure 5.1 shows the occurrence in context of the word "information" in Chapter 3. An editor would first examine the list and eliminate all but the entries with bold terms, on the ground that only in the indicated phrases is the word "information" part of a key term.

The editor could then pass to the full text and have each occurrence of the nominated terms emphasized. For example, the occurrence of "information need" indicated in Figure 5.1 by "<<<<" in the text is shown in Figure 5.2, with other occurrences also emphasized. The editor might decide that the first occurrence of "information need" should be included in the index, while the second and third should not, since they occur in the same paragraph. Secondly, the editor might standardize on the form "information need" rather than "information needs". The software tool could easily present the edited document marked up using XML, so that it could be rendered as a hypertext.

Information storage and	identify the **information needs** of
called an **information storage and**	performed by **information systems pro-**
single word:"information","storage","retrieval"	**fessionals**
a phrase "information storage and	rely on information which is
system" or "information retrieval"	An **information retrieval system**
set. Find "information and retrieval"	for an **information retrieval system**
of Find "information" intersected with	have enough information available to
Find "information or retrieval"	the relevant information
of Find "information" union the	A text-based **information retrieval system**
Find "information and not	finding useful information
of Find "information" with any	An information retrieval system
of finding information illustrated in	to an **information retrieval technology**
a user's **information need**, as	the user's **information need**
to find information is considered	query for "information management" (the
for some information for some	feeding of information as an
to their **information need** meet	about **management information systems**
the **information need** meet	and
the **information need**, but	a user's **information need** can
get some information that can	Suppose one's **information need** is <<<<
the user's **information need**, then	identify different **information needs** derived
query Find "information" and "retrieval"	to an **information need** not
the phrase "information given to	search for information on making
retrieval of information", thereby reducing	number of **information sources** making
to find information about design	An **information space** designed to
obtain such information might contain	to support **information retrieval** have

Figure 5.1 Occurrence in context of word "information" in Chapter 3.

Note, by the way, that Figure 5.1 shows that the term "information retrieval problem" does not occur as such in the text of Chapter 3. The editor would need to decide which part of the chapter the cross-reference hyperlink would lead to.

We can conclude that although it is not presently possible for a collection of documents to be automatically turned into hypertext, there is considerable promise for semi-automated methods, perhaps implemented as XML editors with substantial but simple text processing capability.

Concept map approach for link design

Hypertexts are often large documents, created either by teams of people or by one person over an extended period of time (there are hundreds of people working on the *Encyclopedia Brittanica*, for example). A user, however, sees the document as a single object and has the problem of making sense of it. To the user, therefore, there is a virtue in uniformity. If a certain sort of thing is encountered with links to

User's real need generally requires multiple queries

The foregoing discussion has implicitly assumed that a user's **information need** can be expressed in a single query, and the need can be satisfied from its response. In practice, this is rarely the case. Suppose one's **information need** is to support a paper outlining the possible impact of very low cost, high bandwidth communication technology on the accessibility of medical services in remote communities. If a paper on this topic already existed, it would not need to be written, so a query looking for that specific topic would not be expected to produce any useful material. The researcher would look for different aspects of the topic with different queries: one to identify trends in technology, one to identify trends in telecommunications company tariffs, and probably several to identify different **information needs** derived from particular medical problems of particular remote communities.

Alternatively, one might wish to find out the state of the art in Web-based search engines. A query on a bibliographic database might identify a few items of interest, some of which could be quite old – say, the Harvest project at the University of Colorado. The user might then look for other work by the key authors, other papers referring to the project, and also find the Web pages of the Department having done the work and the agency supporting it to look for other related material.

Figure 5.2 Occurrences of "information need" in the full text of Chapter 3.

particular other sorts of things, the user is more comfortable if this pattern of links always occurs.

Achieving uniformity among a team is difficult, since each person sees and interprets things differently. Uniformity within the work of a single person over time is even difficult, since the way a person interprets the world changes with experience, if not mood. (Remember the principle of uncertainty.) Uniformity therefore requires organizational effort. Organizational effort is often made more effective by the use of tools, sometimes computer based.

The concept map is a particular tool which can help in achieving uniformity within many hypertext projects. A hyperdocument is generally a collection of a large number of objects of a small number of structural types. We can think of it as a whole made as an assembly of parts of various types. In the discussion on structural links above we saw semantic pages, auxiliary pages, figures, references and so on. If at an early stage in the design an agreement is made as to which types of objects are to be connected to which, then a record of this agreement can be consulted by the people designing each object, reminding them which links it should support.

One way to record this agreement is by a *concept map*, which shows the object types as named ovals and the relationships as possibly named arrows. The concept map is closely related to the conceptual modelling tools common in information systems design, such as the entity-relationship diagram. A sample (structural) concept map is shown in Figure 5.3. There are many conventions used in concept maps. In the figure, there is a hyperlink associated with each line in the direction of the arrow. Lines without arrows indicate parts of an object. Using these conventions, we can interpret the figure to say: "A section contains keywords and citations. A

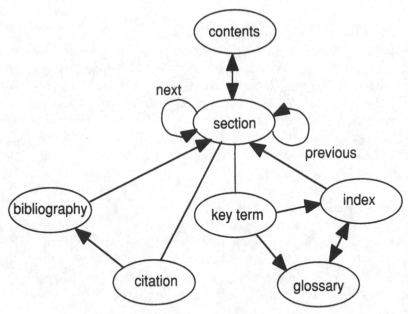

Figure 5.3 A sample structural concept map.

keyword is linked to both its glossary entry and its index entry. An index entry is linked to the section in which its keyword appears, and to its glossary entry, while a glossary entry is linked to the index entry".

A structural concept map is not specific to a particular hyperdocument, but describes a whole class of such documents – say, technical documents. The structural links can be prescribed and implemented in generic software used to build many such systems.

Links that depend on the content of the document rather than its structure are called semantic links. A hyperdocument would have the structural links specified for a document of its class, plus semantic links specific to its own content. Some hyperdocuments are very similar to databases, for example a university catalogue. The catalogue contains thousands of structurally identical subject descriptions and hundreds of structurally identical rules for degree programmes. There are links among subjects, staff members, course rules, departments and so on. The concept map for this kind of document can be a conceptual schema as would be developed for a database application, using for example an entity-relationship diagram. Systems like this are often implemented as renderings of the primary information stored in a database of some kind, often created on the fly.

Other hyperdocuments are structurally less regular than a database, but contain multiple descriptions of similar information. For example, suppose we are constructing a hyperdocument as a compendium of current practice in a broad field of engineering. There would be pages covering current practices, the problems to be solved, the technological platforms on which current practice is based, the historical antecedents of current practice, the changing problem environment, engineering, social and economic constraints, forecast technological developments and consequent possible future practice. A semantic concept map such as in Figure 5.4 would help to

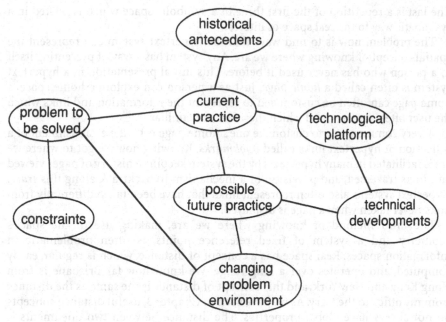

Figure 5.4 Semantic concept map.

ensure that instances of each topic were consistently linked to instances of other topics, even though the topic instances were scattered unevenly throughout the document (there might be only one article on forecast technological developments, say, and the problem environment might be touched on in several articles).

Navigation

We are taking a collection of documents to be an information space, and considering that a user accesses information by passing from document to document via hypertext links. Thinking in terms of a space leads to a spatial metaphor for access, bringing with it a number of familiar concepts for which we can make analogues.

1. One important concept is that of knowing where we are. If we don't know where we are, then even if we know where we want to get to, we don't know how to proceed. This is what being lost means. In real space, knowing where we are has several meanings: We are familiar with a point and its surroundings – we are 'at home' there.
2. We know how we got where we are from somewhere familiar – the Hansel and Gretel solution. (In the fairy tale, Hansel and Gretel can find their way out of the forest because they leave a trail of pebbles.)
3. We have a good idea of the geometry of the region, and a method of relating to fixed reference points – the SatNav solution. (Ocean-going yachts keep their position by reference to a satellite, even though they may be in unfamiliar waters.)
4. We have a map, and know how our location is represented on the map.

The last is a repetition of the first three in a symbolic space which is related in a systematic way to the real space of interest.

The problem now is to find ways that the hypertext system can represent the spatial concept of knowing where we are. Any system has a way of presenting itself to a person who has never used it before. This initial presentation in a hypertext system is often called a *home page*. Just as a person can explore a home space, a home page can often be customized to represent the information and links which the user often accesses, and which therefore are familiar.

A very common way to customize one's home page is by the establishment of a collection of hypertext links called *bookmarks*. Knowing how we got to where we are is facilitated in many hypertexts by the system keeping a history of pages viewed and links traversed, and provision of a mechanism to backtrack along this *trace*. Hypertext systems also often represent links that have been taken differently from links never taken when a page is displayed.

The third method of knowing where we are, making use of the space's geometry and a system of fixed reference points, is often problematic in information spaces. Real space has a concept of distance, which is regular, easily computed, and operates over a long range. We know how far Brisbane is from Hong Kong and New York, and this concept of distance is the same as the distance from my office to the ferry. As we have seen in Chapter 3, useful distance concepts do not always have global properties. The distance between two documents is usefully defined if they are close, but most documents are infinitely far apart (their nearness measure based on shared significant terms is zero). Some information spaces have geometrical properties like real space, and these will be discussed in Chapter 13.

We began the chapter with the concept of map, which is widely used in hypertexts. These maps are representations of the whole of the information space, with some way of indicating where the page being viewed is on the map. This problem is treated in much more detail in Chapter 9.

The first two methods lead to the concept of explored space, while the third and fourth methods involve a representation of the entire space. If we have the concepts of an entire space and explored space, then we can derive a concept of unexplored space (the entire space less the explored space). If the map of the space is annotated to show the explored parts, then the unannotated parts are a representation of the unexplored parts.

How the annotations are represented depends greatly on how the space is represented, which is covered in more detail in Chapters 9 and 13, so we will return to this issue in those places. However, we have considered some representations in this chapter – the hierarchical table of contents (often conceived of as a site map), and the index (or as we will see in Chapter 12 the subject terms or hierarchical thesaurus). In both cases the hypertext system knows which pages are related to which auxiliary objects, so the auxiliary objects associated with visited pages can be annotated by, say, colour or shading. Where the auxiliary objects are aggregated hierarchically, the higher levels can be coloured or shaded in proportion to the proportion of lower-level objects visited.

Finally, in some spaces it is difficult to represent the entire space, but possible to represent a region around a known point. Densely and multi-linked networked structures have this problem – it is notoriously difficult to get an overview of a large entity-relationship diagram in information systems design, or a large and complex hypertext which does not have a primarily hierarchical structure. Both of these examples support a local view – it is possible to see the page of the diagram surrounding a particular node, for instance, or a graph representation of the hypertext pages surrounding a given page. The visited nodes can be distinguished from the unvisited in this view.

This simple regional representation gives a uniform view of a local area, then abruptly hides more distant parts from view. A smoother representation of locality called a fish-eye view gives a detailed view of the nodes immediately connected to the base node, then progressively less detailed views of more distant nodes, with the representation fading into noise in the far distance. An example of a fish-eye view is given in Figure 5.5, which represents a portion of a hypertext supporting a university subject.

The subject home page is the node *home*. The node $W \times W$ is a week-by-week list of lecture topics, while the *L*s and *T*s are pages concerning particular lectures or tutorials respectively. *Ford* and *Bush* are two required readings, while the other nodes connected to home are other aspects of the subject. The two furthest nodes attached to clippings are optional readings. The viewer is at the page *L4*, having reached it from *home* via $W \times W$ and *T4*. The nodes visited are shown in bold, and the links taken are in thicker lines than the links not taken.

This view gives a good representation of where the user has been and where in the local region it is possible to go without saying much about what is more than a few links away.

One property of real space which has not had much of a part in the metaphor of information space is direction. We have used the concept of forward and back when describing the sequence of steps in a path or a guided tour, but have not used other directional concepts like up and down, left and right, or the compass directions. In order to have directions, a space needs to have a geometry rich enough to support the concept of multiple dimensions. A hypertext seen as a collection of pages and links is hard to see as more than a tangle of one-dimensional paths. A user visiting

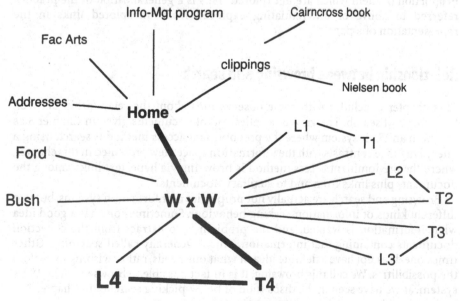

Figure 5.5 Example of fish-eye view.

a particular page has a possibly large number of links to choose from, but each is simply a different way of going forward.

A collection of links is a collection of links, but it sometimes makes sense to differentiate among them. Consider the hypertext represented in Figure 5.5. The page *week-by-week* has a large number of links – there are links to each of 15 lectures and 15 tutorials to begin with. There are about 50 required or optional readings, some of which are books, some articles and some Web sites. A user is faced with nearly 100 choices from that page.

In describing the page, the links have been referred to as belonging to types – lecture notes, tutorials, required or optional readings, books, articles, Web pages. It might be advantageous for the user to be able to request the hypertext system to visualize only the links of a certain type – all lecture notes at one time, all optional articles at another.

Not only might we want to limit the links visible from a given page to those of a certain type to reduce the complexity of choice at a point, but we might want the fish-eye view to show only links of certain types, to reduce the complexity of the neighbourhood shown.

If we have designed the hypertext using the concept map approach, we have a natural way to assign types to links, either structural or semantic, and also to pages. For example, in a hypertext designed using the concept map of Figure 5.4 a page of type *current practice* would support semantic links of types *technological platform, problem to be solved, historical antecedent* and *possible future practice*, as well as the structural link types indicated in Figure 5.3.

Types of links have many of the properties associated with dimensions so we can get some use of the direction aspect of the spatial metaphor. We might think of a selection of links of type *lecture note* as a view of the page in the *lecture note* direction, say. Types as dimensions or directions can give more richness to the concepts of explored and unexplored spaces. The hypertext system might show us what types of links are available from a page, annotated with a representation of the proportion of them which are unexplored. This is a generalization of the practice referred to above of differentiating explored from unexplored links in the representation of a page.

Relationship between browsing and search

This chapter concludes with some observations about the relationship between browsing and search. The view of a collection of documents given in Chapter 3 as stored in an ISAR system where the predominant access method is search using a query language contrasts with the information space view presented in this chapter, where the predominant access method is browsing via hypertext links among the documents plus links from and to auxiliary documents.

Browsing and search are actually not properties of information systems, but two different kinds of information-seeking behaviour. Sometimes one has a good idea what information is wanted, and the problem is to extract from the collection documents containing that information. This is generally called searching. Other times one does not have a definite idea of what one needs, rather wishing to explore the possibilities. We call this browsing. It is in fact possible to browse with an ISAR system, as we have seen in the discussion of berry-picking searches in Chapter 3. It is also possible to search in a hypertext system.

For the sake of clarity, consider a hypertext system without the auxiliary data structures. A large system might have ten million (10^7) pages. Assume that each page has hyperlinks to ten other pages. Then starting from any page, one should be able to expect to reach any other page in about seven steps. Seven steps is not an excessive number – it can take four steps to get from one function to another in a large package like Microsoft Word.

However, some experimentation with large hypertexts and some reflection should convince the reader that it is in fact very difficult to get from one arbitrary page to another. Even though seven steps should be sufficient, it is generally very difficult to tell which link to take at any stage. The difficulty is essentially the problem with the third method of navigation referred to above – the lack of a large-scale geometry and reference points. Driving around a vaguely familiar city, one can often reach the neighbourhood of a destination by dead reckoning. If we start out 10 km southeast of the city centre and know that our destination is about 15 km northwest of the city centre, keeping to main roads going generally north and west can get us close enough to ask directions. In a hypertext there is generally no universal concepts of direction, only local neighbourhoods. There is rarely even a distinction in importance of links corresponding to the concept of a main road.

In practice a well-designed large hypertext has a system of auxiliary information structures such as we have described in this chapter giving a large-scale topology to the site. The point is that hypertext per se does not lend itself to long-distance navigation. Consequently in situations such as in the World Wide Web, where (as we will see in the next chapter) large-scale structures are absent, it is difficult to navigate given only the hypertext links.

The query system provides a universal spatial structure complementing the hypertext neighbourhoods. We have seen in Chapter 4 that the descriptors attached to the catalogue entries of each page are global concepts. A query in a boolean query language identifies a subset of pages. Expanding or restricting the query makes the subset larger or smaller. Assuming a reasonable precision and recall, a query takes us directly to a region of the information space possibly very distant from where we started. It is possible to see the auxiliary data structures often found in hypertext systems as ways of performing queries.

Of course, as we have seen in Chapter 3 the query mechanism has a weak concept of locality – it has a universal spatial structure but the neighbourhoods are not satisfactory. They are either very coarse (boolean query language) or semantically heterogeneous (document nearness query languages). Hypertext has neighbourhoods but weak universal structure. If we are searching, a query followed by navigation through a few hypertext links is often successful. If we are browsing, a query can often define regions in which hypertext navigation is very informative, and a hypertext trip around a neighbourhood can often suggest valuable query terms.

Why can this work?

A bare hypertext is simply a collection of pages linked together. Much of the message of this chapter is that if the collection of pages is large, the user needs considerable auxiliary data structures in order to support finding their way around the collection. Further, the chapter has argued that a large hypertext is better understood if it has a uniformity of design, advocating a conceptual map approach to achieve this.

Uniformity of design requires a discipline among the people doing the design, which implies some sort of organization. Auxiliary data structures require design and implementation, which must be done by someone with an overall view of the information, which also implies organization. Finally, the auxiliary data structures must be stored somewhere, which implies the existence of some sort of organization to provide the storage space and access. A well-organized hypertext requires an organization. The next chapter examines what happens when this organization is missing on the World Wide Web.

Key concepts

A **hypertext** system is one type of **information space**, which consists of a collection of **pages** connected by **links**. One way to establish a set of **types** for pages and links is to use a **concept map**. A sequence of links taken by a user is called a **path**. A user following a path can become **lost in hyperspace**. One way to prevent this is to allow the user to establish **bookmarks** at significant points in the path.

Further reading

There are a large number of books on hypertext and multimedia design, both for CD-ROM and single Web site presentation. Concept mapping or mind mapping is discussed in many guides to study and writing essays. The more formal concepts of information systems modelling using the entity-relationship model are covered in most introductory information systems design texts, in particular Elmasri and Navathe (2000).

Major formative exercise

As in the major formative exercise from Chapter 3, pretend that you are going to write a 3000-word essay on some topic, which is sufficiently narrow to discuss, at most, five items of information. The items of information should be a mixture of pages from hypertext sources, either a CD-ROM hypertext or World Wide Web pages in large Web sites, possibly reached through search engines.

1. Make a detailed record of your experiences, including:
   a. links traversed;
   b. the queries you made (text of the query, system query made on);
   c. description of the results (this can be summarized);
   d. the proportion of the results relevant to your topic (precision) at each stage;
   e. hypertext links taken, and links which were available but not traversed.
2. Present the topic for your pretend paper in a sentence or two. Comment on how the topic evolved during your search.
3. Make a concept map of a Web site (or other large hypertext) including the item(s) you retrieved. The map should contain at least ten concepts, including

some of the links not taken. Note how many links and/or pages are associated with each concept.

4. How confident are you that you have all the information relevant to your (remember quite narrow) topic that is contained in the sources you used (high recall)? Justify your opinion using evidence from your particular experience.

Your report should give general conclusions **based on your specific experiments**.

Tutorial exercises

1. Consider a possible hypertext implementation of this text, where the objects are the lowest-level titled subsections. What subsections might be linked to the subsection "Relationship between browsing and search" in Chapter 5? How did you make that decision?

2. Extend the implementation to include the entire bibliography for the course, assuming that all material was available in full text. What sorts of links would you make? How would you establish them? What would the high-level organization look like?

3. Consider the entries in the telephone *Yellow Pages* for restaurants. If they were to be included in a hypertext system, possibly a Web site, what sorts of links might we establish among themselves and with other kinds of object? Make a concept map of the information domain.

Open discussion question

What problems would you envisage in keeping the hypertext of Question 3 current? How might you overcome them?

6 World Wide Web

What happens when you move out of a hypertext into the World Wide Web? This chapter introduces concepts of resource discovery, third-party information structures and information flows.

How is the Web different from a library catalogue or a CD-ROM?

The chief organizing principle for the World Wide Web is hypertext. The essential feature which makes the Web possible is the ability to make hypertext links across computer systems using the Hypertext Transfer Protocol (HTTP) and the Universal Resource Locator (URL). If the Web is a hypertext, then one might ask what must be said about the Web that is not implicit in the previous chapter. How is the Web different?

First of all, the Web is *big*. Think of a large hypertext, for example the *Encyclopedia Brittanica*. The *Brittanica* in book form contains about 35,000 pages, each of which has perhaps ten topics, so the CD-ROM version has perhaps 350,000 pages. The Web has between 10 million and 100 million sites. The size of the sites varies enormously, but the average might be somewhere between 100 and 1000 pages. This means that the Web includes between 1 billion and 100 billion pages. The Web is between a factor of 1000 and 100,000 bigger than the largest hypertexts.

Second, the Web is *varied*. A hypertext publication contains a few types of objects all aimed at a similar audience – say pages of text, pictures, sound bites, video clips and perhaps a few interactive simulation programs all aimed at high-school students. The Web contains dozens of media types aimed at an enormous variety of audiences. There are universal sites like that run by Microsoft or CNN. There are personal sites intended for a student to share things with friends or a grandparent to publish photographs of grandchildren. Businesses have sites aimed at their often quite specialized clientele. All sorts of interest groups have sites aimed largely at communication among group members.

Not only is there a variety in media type and audience, Web sites differ in granularity. They range from a ten-page personal site to library catalogues and online bookshops with millions of records in their databases. The *Encyclopedia Brittanica* is a Web site. They differ in authority. Medical sites run from Medline, which publishes refereed research papers, through advisory sites run by Departments of Health, product information sites run by pharmaceutical companies, information and support sites run by major support groups like the Arthritis Foundation, to chat sites and opinion sites run by small groups of people of all sorts. They differ in quality, price and quality of service.

Size and variety are very large differences, as we will see below, but the most significant difference is organizational context. A large hypertext like the *Brittanica* is designed, built and maintained by a single organization. As we saw in the previous chapter, a single organizational context is necessary for there to be a coherent design and uniformity of presentation, and is also necessary to build, maintain and house the auxiliary information structures which contribute so much to the practical usefulness of a hypertext system as an information space. The Web, on the other hand, has millions of organizational contexts. There is no one to enforce coherence of design or uniformity of presentation, and no one to build, maintain or house the auxiliary information structures. At bottom, the Web is a pure hypertext. The main point of this chapter is to show how the limitations of a pure hypertext can be overcome in an organizationally heterogeneous environment, especially when the size and variety of the Web is taken into account.

An obvious consequence of organizational heterogeneity is that one often attempts a hypertext link only to be informed that the site requested no longer exists. Broken links are a result of incomplete maintenance of a hypertext – a page is removed without all the links pointing to it having been removed. In a hypertext such as that considered in the last chapter, when a page is removed there will exist mechanisms to identify all the links to it, so that the change can be propagated. These mechanisms depend on a unitary organizational environment responsible for the integrity of the hypertext. On the Web, once a page has been published, others all over the world can make links to it without informing the publisher. There is neither a mechanism nor a place for a mechanism to reliably propagate a page's deletion.

The Web has tens of billions of pages, but only tens of millions of publishers, so may be thought of as a sort of society of hypertexts of varying sizes. A single Web site has the character of a hypertext of the previous chapter, some of whose links are external. Maintenance of the external links can be seen as the responsibility of the site's publisher, and the easiest way to do this is by periodically traversing them, possibly automatically. This method is called *polling*.

On the other hand, many sites are intended to be repeatedly accessed by their clientele, either by being bookmarked in individual Web browsers or by having links incorporated in other sites. We might think of these sites as *strongly present* on the Web. They are somewhat like major landmarks in physical space.

The search engines are examples of strongly present sites. These sites have a strong business reason to maintain the integrity of their incoming pointers. However, the rapid evolution of the Web and of their own presence on it requires frequent revisions of their own sites, sometimes at the highest level of the URL address. They face the engineering problem of changing their site in the face of a strong legacy presence.

A simple way to maintain the legacy presence is to maintain an archive site which contains targets for the intended-to-bookmark pages, either redirecting the incoming link to a comparable page in the current site or explaining the demise of the previous page with an apology. An example of such a message from a major Web presence having undergone significant reorganization is:

This is the DSTC Archive Site, your file was not found at the new DSTC Web Site. To access the new DSTC Web Site please choose: Home Page Sitemap Search.

Finally, there are emerging a large number of third-party value-added Web sites, which repackage and integrate information from other sites. These sorts of sites are discussed in more detail later in this chapter. Especially if these sites are revenue-generating, which increasingly they are, and if their input sites are also revenue-generating, it pays both sites to give up some of their autonomy and undertake a contract to maintain a link and the service it carries for a defined period. It then becomes the responsibility of the target of the link to notify the origin of any change with sufficient notice to update the origin's site so that the service can be maintained. There are many possible protocols for doing this.

In summary, well-organized Web sites maintain external links by polling; strongly present Web sites maintain incoming links by archiving; and sites linked by business contract maintain links cooperatively. Sites which are poorly organized, weakly present and with adventitious connections leave the Web strewn with broken links. This is a most obvious way the Web manifests itself as a society rather than an organization.

Resource discovery

The previous chapter concluded with the observation that searching in a large hypertext was very difficult, since the spatial structure provided by hypertext links is very local. We speculated there that in a large hypertext two pages might be perhaps seven hyperlinks apart, which especially in the absence of any global concept of direction made long-distance navigation very difficult. This concept of the average distance between any two pages was formalized in a study in 1999 as the *width* of a hypertext. Since the Web is very much larger than a single hypertext, we would expect that the width would be greater also. That study estimated the width of the Web as 19 – on average it takes 19 clicks to get from one randomly chosen page to another. If a hypertext with a width of seven is difficult to navigate in, how much worse is the Web with a width of 19!

We saw that searching, finding information that one already knows something about, requires a global view of the information space. Hypertexts solve this problem with auxiliary integrative information structures, but we have seen that integrative information structures require someone to build and maintain them, and also require someone to house them, and that the World Wide Web lacks the organization needed. Providing facilities to enable users to search the Web is therefore a significant engineering problem, which has been given the name *resource discovery*.

For the first few years of the Web's development, that is to say, up until just about the early 2000s, the chief resource discovery tool was the search engine. A search engine treats the Web as a text database along the lines described in Chapter 3. It is operated as a Web site, one among the others, which catalogues pages from other Web sites in a database, and provides a query interface to its users. There are thousands of search engines, including tens of large ones.

The Web has taken its principle of organization from the hypertext. The search engines provide an integrative mechanism by treating the Web as an ISAR system. We have seen above that the hypertext organization breaks down because the Web is very much larger and much more varied than a hypertext, and most tellingly, the Web is a society rather than an organization. These same factors make the view of the Web as an ISAR system ultimately untenable, and are therefore responsible for the weaknesses of the search engine concept.

Size is a problem, but of itself not decisive. The Web has say 10^{11} pages. A large library, say the Library of Congress in the US, has 10^7 volumes, each of which has somewhat fewer than 10^3 pages, for a total of 10^{10} pages. The Web, seen as a collection of pages, is therefore not much bigger than the largest collections of information managed by ISAR systems. (The fact that the Library of Congress is itself a single Web site is a different problem, discussed below.) Especially, we have noted that the number of sites is comparable to the number of volumes.

Variety is also a problem, but similarly not decisive. If an ISAR system can keep track of tens of millions of documents by millions of authors and hundreds of thousands of descriptor terms, it can surely cope with thousands of document types instead of the tens of document types encountered in a library.

The critical factor in the breakdown of the ISAR view of the Web is lack of overall organization. We have seen in Figure 2.1 that an ISAR system is built on a database of surrogates for the documents of interest. The search engines are the same in this respect. The problem is how to create the surrogates.

A library, or a publisher of a CD-ROM, is an organization with a defined purpose, intending to serve a specific clientele in a well-understood set of tasks.

- A university library is intended to provide materials for teaching the curriculum and for research into a range of areas – the research being in a limited variety of genres.
- A parliamentary library is intended to provide materials for policy analysis.
- A criminal information system is intended to store materials related to a specific set of criminal investigations.

Organizational information sources limit the acquisition of new material to that which is relevant to their purposes.

Control of the material and knowledge of the uses to which it is to be put is essential to the success of the cataloguing effort as described in Chapter 4. A controlled vocabulary is designed, maintained and applied by the organization (often outsourced to a major deposit library such as the Library of Congress or the British Library). Even where the descriptors are computed from the text of the materials catalogued, the text is controlled in some way. A bibliographic system which extracts descriptors from document abstracts must take what it is given. What is given to it, however, is subject to many controls. Journals have standards for abstracts, and the refereeing process enforces them. Moreover, the bibliographic system limits its content to a limited collection of journals of limited types (there may be thousands of journals of hundreds of types, but this is far less variety than found on the Web). Newspaper articles are written to editorial standards, and cover a limited (admittedly large) range of topics.

Even so, the cataloguing process is not entirely successful. It is well known that students must learn research skills in order to effectively use their library resources, and that there is a substantial return on a fairly large amount of training (many libraries offer courses lasting an hour or more).

Most search engines construct their catalogue entries automatically, as described in Chapter 4, from the Web pages themselves. They also construct brief precis from the pages, using methods described in Chapter 7. The process of constructing a catalogue is the same as any ISAR system. The difference is that there is more variety and much more uncertainty (Chapter 4). The fact that an organization builds a collection to serve specific purposes reduces uncertainty –

there is a limited range of queries – and the provision of training reduces uncertainty even further. The organization's control of the cataloguing process decreases variety. In a bibliographic database, the broad similarity of culture in scientific journal publication decreases variety. Even the fact that a journal provides an identifiable block of text as an abstract requires organizational discipline.

The pages indexed by the search engine come from millions of sources and tens of thousands of genres, so the variety is enormous. The search engine is intended to be used by anyone who wants to find anything on the Web, so the uncertainty is enormous. This is why the precision and recall of Web searching is often very bad.

Of course the situation is not uniformly so bad as painted in the previous paragraph. Some terms are names of widely known products or organizations, so there is appropriate variety and not too much uncertainty. Search engines are quite good at finding information about such things. Further, Web page designers can take deliberate steps to mesh in with the search engines.

Designers can make the variety/uncertainty less of a problem, if the designer gives thought to how the desired market is likely to describe what is on offer, and include those terms on the page so that they will be included in the search engines' catalogues. This is the same process as that used by organizations to catalogue their collections in ISAR systems, as described generally in Chapter 4 and more specifically in Chapter 12.

However, designers can also make the problem much worse. Some Web sites identify the most commonly used search terms, and include them in their pages regardless of the relevance of their page to the term, hoping that their site will be in nearly everyone's result set. This practice, called *spamming*, is strongly discouraged by the search engine organizations, and their cataloguing algorithms attempt to detect it. A page deemed to be spamming will not be catalogued, but of course the spammers continually try to outwit the cataloguing algorithms.

Some search engines attempt to exercise more control over what they index. Yahoo! is the paradigmatic example of this type. (Yahoo! is strictly not a search engine, but a classification system.) Its organization is based on a classification scheme such as that described in Chapters 8 and 11. However, the effort to control the description of the sites indexed is considerable, so that for reasons of cost only a small minority of sites are included in their ISAR. They retain a more reliable ISAR giving access to the most prominent sites at the cost of exhaustivity.

A further factor in the increase of variety and uncertainty is the spread of the Web to languages other than English. More and more content is being provided in a wide variety of languages. There is also a rising number of automatic translation services, so that much of the Web can be accessed in multiple languages. However, seeing a site through perhaps two levels of translation (say Spanish to English followed by English to Chinese) means that the user has to contend not only with the variety of ways the content providers can describe their sites, but also with the vagaries of the translation programs.

Managing a large and increasing variety and uncertainty is a major problem for the search engines, but a more serious problem is that more and more of the Web is invisible to them. Search engines rely on the Web being a society of interlinked hypertexts. Their cataloguing programs traverse a site starting from a home page and systematically visit other pages on the site by following links.

More and more sites are not hypertexts, but ISARs in themselves. There are thousands of library catalogues, hundreds of bibliographic sites, and many online bookshops, CD shops, and databases devoted to a wide variety of topics. These sites have only a few Web pages with hardly any content, whose function is to provide a user interface to the underlying ISAR system. The search engine cannot access the content of the sites, therefore cannot catalogue it, so the user of the search engine is not directed to that content. We might call these sites *invisible sites*.

This section has shown that the large-scale order in the Web has heretofore been provided by search engines using ISAR technology, but that the Web is evolving beyond search engine capability. Some speculations as to the future of Web organization are given below.

The emerging order (sort of) in Web information space

We have seen above that the Web differs from a hypertext most fundamentally in that a hypertext is produced by an organization which can also develop the auxiliary data structures needed to support the large-scale structure of the information space, and which can provide a place for these information structures to reside. The Web has no such organization, and so lacks the large-scale information structures.

However, we have also seen that it is possible to develop sites (the search engines) whose purpose is to organize other sites, thereby providing a semblance of the large-scale structure needed for the Web to work well as an information space. There are a number of other approaches to building sites adding structure to the Web, which are rapidly evolving in the early 2000s. The terminology for such sites is also in rapid evolution.

Portals

Search engines were originally expected to be a universal interface to the content of the Web. As the Web has increased in size and variety, this expectation has failed in at least three ways. The first two, described above, are the *principle of uncertainty* – someone seeking information may describe it differently from those providing it – and the *invisible Web* – that many database-driven sites are not indexed by the search engine. Added to these two is what might be called the *principle of ignorance* – if one doesn't know what is out there, one doesn't think to look for it. One must know a fair bit about the information one is seeking to make successful use of an ISAR system.

As we have also seen, some search engines, notably Yahoo!, work on the principle of a classification system rather than a bare query interface to an ISAR system. The classification system provides a view of the Web similar to a table of contents in a hypertext, or to a hierarchical thesaurus as described in Chapter 12. This solves the problem of uncertainty, since the display of the classification system tells the seeker how the providers describe their information. It solves the problem of invisibility to a degree by providing slots for the database-driven sites to be visible (the sites only, not their content – see below). It solves the problem of ignorance by presenting the range of possibilities to the user.

Classification-based search engines have two other problems. First, alluded to above, is a lack of exhaustivity. It is expensive to classify sites, so the coverage is low. Second, specific popular sites are often buried deep in the hierarchy, so are difficult to find. The first problem is generally solved by provision of access to one or more of the ISAR search engines. The second problem is solved in practice by individual users bookmarking the sites to which they require frequent access. A user's set of bookmarks is a fairly complete, if very idiosyncratic, view of the Web for that person's purposes.

Portal sites are attempts to universalize the experienced users' collections of bookmarks. They combine a classified view of the Web with ISAR search engines and a collection of specific sites intended to cover all the popular specific choices. Portal sites are generally mounted by the major search engines or major content packagers. They are intended to serve as everyone's home page – the starting point for Web access (hence the name "portal'), and they succeed to a degree.

They run afoul of the principle of variety, however. It is reported that they work well for new users, since they overcome the principle of ignorance, but experienced users of the Web have access to so much material that their own bookmark collection very quickly becomes superior to the portal site. There is so much variety on the Web, and so many users with such different information requirements that a single view of the Web, even if the portal can be personalized, cannot work.

Vertical portals

We have seen that classification-based views of the Web tend to be incomplete due to the cost of indexing, and that the attempt to create universal portals is likely to be successful only for some users some of the time. Other sites have taken the approach that if we are doomed to an incomplete view of everything, then why not try to control the incompleteness and get a complete view of a limited domain.

There are a large number of these, what are often called *vertical portal* sites, in all sorts of domains: particular industries like mining; particular problem areas like recruiting in a particular industry, health advice and information or second-hand books; particular lifestyle domains such as philosophical-literary discourse, adventure tourism or guides to the attractions of and services available in particular tourism destination areas. These sites attempt to improve precision and recall by drastically reducing variety (coverage of a limited collection of sites) and also uncertainty (the sites are semantically related, and tend to use similar terminology to the expected users).

As these sites are domain-specific they can attain a more complete coverage, since their creators can feasibly perform sufficient research to do so, and also they can establish contractual relationships with content sites to ensure currency. Since the domains are limited and the sites are aimed at specific audiences, there are also many possibilities for funding the sites, including targeted advertising and various revenue-sharing arrangements. It is common for a virtual vertical portal to be formed by a voluntary association of related sites which contain links among themselves – a so-called *Web ring*.

Vertical portals are much smaller than universal portals – typically tens or hundreds of sites, rarely more than thousands; while the universal portals attempt to cover at least millions if not the entire Web. Vertical portals are not thought of as home pages, but as indexes of particular domains.

There may be a trend for universal portals to recognize their incompleteness and to reduce their coverage to tens or hundreds of thousands of sites of potential interest to defined market segments. They are calling their top level categories channels, and associating particular collections of keywords to channels for use by agreement with the content sites classified into those channels. There is also a move to agreements between more universal portals and vertical portals for participation in and sometimes control of particular channels.

How might this order develop?

The Web is evolving rapidly, so that it is unwise to be definitive about what the Web looks like, and impossible to foresee what might happen in the next few years. We are not without resources, though. It is not as if the problems being solved on the Web are problems which have never before been faced by our society.

We are concerned with how the Web can be organized so that people can find the information they need on it. People have been finding the information they need long before the introduction of the Web, in what we will term the real world (as distinguished from the virtual world of the Web). The Web is a very large information space, but the real world is a very much larger information space – in particular it includes all the content on the Web. We can get some ideas by looking at how the real world is organized to help people find things. It is a reasonable bet that most of them will have analogues on the Web.

Some of the information structures in the real world are organized around similar kinds of resources:

- the yellow pages, organized around classes of goods and services providers;
- product catalogues, organized around product types;
- airline timetables, organized around origin, destination and time;
- university catalogues, organized around discipline, subdiscipline and level of difficulty;
- organization charts, organized around reporting structures; travel guides, organized around tourist attractions, hotels, restaurants, etc;
- cookbooks, organized around type of dish and ingredients.

These sorts of information sources are developed and presented from the point of view of the supplier of the resources.

However, we have seen in Chapter 3 that people seeking information very often want several items from several sources (the berry-picking metaphor). People seek information to assist them in some activity or making some decision.

- A student planning a university study programme will choose several courses from different disciplines, consulting not only the course descriptions but also the programme rules and the timetable.
- Planning a trip can involve airline schedules, tourist attractions, rail timetables, hotels, car hire, immunization, travel alerts, visa requirements and many other items of information.
- Planning a meal can involve several recipes for different courses, sources and prices of ingredients, possibly sources of specialized equipment, and weather forecasts.
- A minor home renovation may require sources and prices of a variety of materials, tools, and specialized trades and transport services.

In the real world there are information resources supporting these kinds of complex activities, notably specialized magazines. There are tens of thousands of these – every lifestyle or activity shared by more than a thousand people has at least one magazine for it. Every business role involving purchasing has trade magazines supporting it. There are major publishers who publish only these.

Lifestyle and trade magazines typically contain:

- case studies of activities typically performed by that group, including how-to-do: sources of ideas;
- surveys of a range of products, technologies, or approaches to a problem encountered by that group: criteria for choice;
- advertisements for products and services relevant to that group: specialized entries into a variety of information resources.

They are organized around supporting the activities of a specialized group of people. Each specialized magazine touches many different information sources of the previous type. It makes sense to name these two classes of sources. We will say that the supplier-oriented sources are in *resource space*, while the others are in *behaviour space*.

Categorizing real world information sources into resource space and behaviour space allows us to see that there are resources in between the two spaces.

- Some travel books are in resource space, some in behaviour space; but there are travel bookshops, which sell nothing but both kinds of travel books.
- Supermarkets, historically organized as resource spaces, are introducing recipes and their own food lifestyle magazines.
- Universities, banks, insurance companies and other providers of complex catalogues of services have introduced advisory facilities to assist clients in selecting an appropriate set of units to meet their particular requirements.

Services are in resource space if a complex query requiring a berry-picking approach accesses several unconnected services, or perhaps the same service with several disconnected queries. We can see that many of the existing Web services are in resource space, especially the invisible sites. The search engines appear to tacitly assume that the entire Web is in resource space. Universal portals can also usefully be described as in resource space.

Behaviour space is less well developed. Services are in behaviour space if a complex of information needed for a task or activity can be found from a single site or several richly interconnected sites. Vertical portals are often in behaviour space, being limited to a particular industry or to the facilities of a particular small region. There are also very many individual sites which can be seen as in behaviour space – Star Trek and other fan sites, religious sites, political sites, and all sorts of lifestyle-oriented sites catering for a few thousand or even a few hundred people. Many of these behaviour space sites are not intelligible to people outside their target audiences.

It is likely that with the rise of mechanisms to fund Web sites there will be a proliferation of behaviour space sites and an improvement in their production values and a proliferation of vertical portals. These structures will eventually tie the Web together in a dense network of third party sites, making it much more like a giant hypertext.

Pull and push – advertising

Most use of the Web involves a user seeking information. The concept of *push* – information seeking a user – was introduced in Chapter 4. The term "push" comes from Web applications – the concept in ISAR systems has other names, such as "alerters" or "subscription". In contrast with push, remember that we call the situation where a user seeks information *pull*.

Some sites are regularly updated (e.g. news sites or discussion sites), so that their users must regularly access the site to keep up-to-date. Many of these sites allow a user to install a filter, so that when an update is posted matching the filter a message is sent notifying the user. Some systems use e-mail for this, others have a direct channel to a screen saver or to a permanent window in the Web browser.

A significant weakness in this style of push is that the user must prepare a description of the information they wish to receive. The most interesting information though is news about something unexpected that might affect the subscriber, either positively or negatively. For example, it is unlikely that anyone in the business community would have been notified about news reports of the introduction of the Web.

In the real world, a significant arena where people seek information and information seeks people is in advertising. Advertising tends to be thought of as the ultimate push – broadcast advertising on television and the mass-circulation press – but there is also a significant pull component. People use classified advertisements to find goods or services they need. There are publications devoted to classified advertisements which people buy when they need something. The specialized magazines are often most useful to their readership for their advertisements.

Informing people about new things they might be interested in is one of the central problems of advertising. Ideally, an advertiser would like an ad to come to the attention of all and only those interested in it. In practice this is impossible, but what is done is to attempt to identify the potential customers by easily obtainable information about them (so-called demographics or profiles). A product with a market limited to people interested in ocean sailing in wooden boats would be advertised in a magazine devoted to that market. However, a product expected to appeal to men with limited ties, moderately high income and significant leisure time would probably also be advertised in a wooden boat ocean sailing magazine, among others.

This suggests another version of push technology, where instead of describing the information they are interested in, users describe themselves. A site then assesses which user profiles are most likely to be interested in a new piece of information, and pushes it to them. This sort of push is becoming fairly common, with computers or Internet service provider connect time being provided free or at a large discount in return for a detailed profile and a willingness to accept a stream of targeted advertisements.

We can expect that the Web will evolve into a more balanced flow of information, with a combination of pull and push, with the push controlled by a combination of descriptions of information and descriptions of people, probably mediated by a rapidly evolving class of program called *agents*.

Agents and the need for standards

In order for two objects to communicate with each other, each must know how to send and receive the messages, and must know how to interpret them. In a distributed application, the methods used to communicate, the structure of the messages, and the behaviour expected on their receipt are all part of the system design. But the system is designed, built and operated by an organization, and the fundamental characteristic of the Web is that no organization subsumes it – we have called the Web a society rather than an organization.

In the Web, the structure needed to exchange and interpret messages is provided by standards, which are agreements among the organizations exchanging messages as to how to do it. There are a large number of standards facilitating the Web.

- Communication protocols which get strings of bits from one place to another.
- Agreements on how to describe the structure of messages contained in the bit strings (the markup languages HTML and XML are the subject of Chapter 7).
- Access protocols which define what is meant by queries and through which a complex series of queries may be performed, for example Z39.50. Z39.50 originated in the library world and is oriented towards remote access to ISAR systems. It provides standard query languages, ability to manipulate result sets, and the ability to formulate alerters which filter the stream of updates to the ISAR system as in the discussion of push technology above.
- Directory standards through which particular people or organizations can be found, for example X-500, which is a white pages style directory organized either by geographical region or organizational subunits.
- Standards by which services can be described, often called metadata, such as Dublin Core.
- Standards describing what the words mean, such as those organized around Z39.50 as standard attribute sets in particular domains, e.g. bibliographic data (BIB-1), government data (GILS), descriptions of museum collections (CIMI), or scientific and technical statistical data (STAS).
- Standards describing particular types of messages which have business meaning (requests for quotation, quotes, orders, invoices, electronics funds transfer), which have their origins in Electronic Data Interchange (EDI).

Metadata, standard attribute sets and EDI are described in more detail in Chapter 12.

The Web, as has been described so far, is essentially a passive medium, responding only to information-seeking activity by users. The exception has been push technology, where the Web behaves as an autonomous information source. However, just as the real world goes on autonomously with respect to any individual person, the potential is there for the Web to be animated by large numbers of programs which make it an active medium with which a person can interact, in much the same way as co-workers interact.

The sorts of programs which are being implemented to make the Web active are often called agents. In order to perform useful work, agents generally rely on standards. There are a large number of these already operating.

- The push subscriptions, alerters etc. described above can be seen as autonomous programs which belong to the subscriber but which reside on the server operating the ISAR system being filtered, using Z39.50 or related standards.

- Web crawlers, which retrieve Web pages as HTML documents and perform some process on them, link from page to page via the HTML link tags. Search engines generally construct their catalogue entries using Web crawlers, but they can be used to search for anything that a program can be written to recognize in a block of HTML text. These of course depend on the HTML standard, and increasingly on XML.

- Shopping bots, which perform comparative shopping for standardized products such as books or CDs. The first wave of these operate by hard-wiring the bot programs to the known retail sites, and processing the HTML forms pages which form the retailers' user interfaces. The future of these sorts of agents lies in interacting with yellow pages style brokers (sometimes called *traders*) and retailers using industry-standard EDI messages, probably represented in XML.

- More sophisticated programs acting as buying or selling agents in auctions. Automated program trading on various financial exchanges is the most developed of these. They rely on message-type standards similar to EDI.

In sum, the Web is evolving very rapidly both in technology and applications. However, the fundamental information science principles through which it is organized are progressing at a much slower rate, and will continue to be applicable for some time.

Key concepts

Finding sources of information on the Internet is called **resource discovery**. The information structures used to assist resource discovery can be either in **resource space** – they are organized so that similar resources are brought together – or in **behaviour space** – they are organized so that resources used in particular behaviour patterns are brought together. A **universal portal** is a site intended to give a user a complete view of the Web. A **vertical portal** is a site which is intended to give a complete view of a limited domain.

Further reading

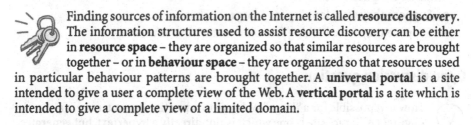

There is a large amount of literature about the World Wide Web, but since the area is so new, and developing so rapidly, the material available is scattered. An early discussion of the issues is given by Lynch (1995). A discussion of how meaning is established by the users of large systems can be found in de Certeau (1984). More information on the various standards can be found by judicious searching on the Web. The concept of width of the web was introduced by Reka et al. (1999).

Formative exercise

Revisit the major formative exercises from Chapters 3 and 5, performing them on the Web.

- How do you combine the search engines and hypertext links to find the information you are seeking?

- What use did you make, or could you make, of universal or vertical portals?
- How do the portals you use manage their relationships with their content sites?
- Classify the key sites you used in resource space, behaviour space, and in between. Is there any difference in the relevance or quality of information from the various types of sites?

Tutorial exercises

1. Consider a Web site devoted to this course.
   - What might it contain?
   - What sorts of people might want to find it using a search engine?
- How should it be catalogued? Would the contents of the site be adequate?
- What would someone wanting material from this Web site be doing that they needed the information?
- What other collections of information would be similar resources to this Web site?
- What problems would you envisage in keeping the site current? How would you overcome them?
2. How would a Web search engine help someone find the CD-ROM (bibliographic) database you are using for the formative assignment from Chapter 3, assuming it was on the public Web?
   - Suppose they were looking for a large database.
   - Suppose they were looking for some material as you were.

Open discussion question

How is it possible for a "large" Web site to advertise its contents via a search engine? (A "large" site is one which is not directly a hypertext, but generates its pages on the fly from database queries.) Examples of such sites include a major library catalogue, CD-ROMs, telephone Yellow and White Pages, classified ad sites, online bookstores like amazon.com or barnesandnoble.com, online CD stores like CD Now.

Consider various possibilities, then speculate as to how they would interact with typical search engine behaviour. How often would a user find particular large sites useful? How much irrelevant traffic would one of these large sites get? Note that amazon.com has a co-search arrangement with Yahoo!, and Barnes and Noble has such an arrangement with Webcrawler.

7 Structured documents: XML

Not all documents are simple. Structured documents have their own set of problems and opportunities, both as objects in a larger information space and as information spaces in themselves. Structure is represented by markup languages, most interestingly XML. XML can be used for a variety of purposes.

Complex objects

Up until this point, we have implicitly considered that the elementary object in our information spaces were simple, small and internally homogeneous. In the information storage and retrieval discussions of Chapters 1–4 we thought of collections of library catalogue entries, abstracts, pictures described by captions, and so on. In the discussion of hypertext of Chapter 5 we looked at complex documents, but thought of them as one object with parts which were simple. The issue was how a complex document like a reference book could be constructed out of simple parts connected by hypertext links and organized with auxiliary information structures.

However, the World Wide Web is so huge that even a very large hypertext like the *Encyclopedia Brittanica* is only a tiny fragment. Many large collections of complex objects are becoming available.

- Many publishers of scholarly journals make available full text of articles. A publisher may have hundreds of journals each with thousands of articles.
- More than 10,000 (in the early 2000s) books are available in full text.
- Large video libraries exist. A single video consists of a complex organization of scenes and shots.
- Large collections of spoken dialogue exist and are increasing (this is one aspect of a video object), which have complex structures.
- A record of a complex business transaction is itself a complex object. These huge databases are beginning to be published in one form or another.

It is useful therefore to take a multi-layered approach, and to think of large collections of complex objects.

There are many ways of representing complex structure, but the method most relevant to the Web is via what is called a "markup language", most particularly XML, to which we devote the remainder of this chapter.

Markup languages

Markup languages originated in the printing industry. Historically, the author of a document to be printed prepared a manuscript containing the content of the work, which was then prepared for printing by people working for the publisher. A manuscript was minimally formatted (originally handwritten), while a printed work had a complex and regular layout. An editor "marked-up" the manuscript with instructions to the typesetter who then put it into the desired format.

Computer-management of text leads to a similar problem. To the computer, a text is a collection of integers of a size depending on the code used. ASCII codes are 8-bit integers, while Unicode codes used for pictographic languages are 16-bit integers. Originally, electronic texts were minimally formatted typescript designed to be viewed on simple devices like teleprinters or character-capable video screens (dumb terminals). For example, Figure 7.1 contains the typescript for the formative exercise from Chapter 6.

Devices have always existed which could render the text in a typeset-like form. (A modern wordprocessor displays it on a screen as it will look when it is printed.) In order to make the text appear in a desired format, it is necessary to insert commands to the rendering device in the text. A possible markup for the typescript in Figure 7.1 is shown in Figure 7.2.

The markup commands shown as lines beginning with "." are schematic. A particular device would have particular command codes to achieve these effects. This technique works well, and is in fact automated by most wordprocessors and all page layout programs. However, it has a major limitation.

We often want to make radical changes in the way a document is laid out. For example, if the text of Figure 7.2 were to ultimately be published as in Figure 7.3, with a 10 point font and multiple columns, the author would probably prefer to work in draft in a larger font and a longer line, which can more easily be viewed on the screen, and which prints in a form that is easier to write corrections on. The markup has to be changed when the document moves out of draft into publishing.

A text marked up with a command markup language as in Figure 7.2 would require massive changes in the markup commands to achieve rendering as in Figure 7.3.

Looking at our example, however, we can see that there are only two text elements in it: a normal paragraph and a list of bullet points. A normal paragraph is always rendered the same way, and so is a list of bullet points. If the text were marked up with a more semantic markup language describing text elements, then

> **Formative exercise**
> Revisit the major formative exercises from Chapters 3 and 5, performing them on the Web. How do you combine the search engines and hypertext links to find the information you are seeking? What use did you make, or could you make, of universal or limited-domain portals? How do the portals you use manage their relationships with their content sites? Classify the key sites you used in resource space, behaviour space, and in between. Is there any difference in the relevance or quality of information from the various types of sites?

Figure 7.1 Typescript.

```
.set page width to 210 mm
.set page length to 297 mm
.set top margin to 2.54 cm
.set bottom margin to 2.54 cm
.set left margin to 3.17 cm
.set right margin to 3.17 cm
.set font to helvetica
.set size to 14 points
Formative exercise
.set font to times
.set size to 12 points
.skip line
Revisit the major formative exercises from Chapters 3
and 5, performing them on the Web.
.return
.skip line
.indent 1 cm
.hanging indent 2 cm
.insert bullet
How do you combine the search engines and hypertext
links to find the information you are seeking?
.return
.skip line
.insert bullet
What use did you make, or could you make, of universal
or limited-domain portals?
.return
.skip line
.insert bullet
How do the portals you use manage their relationships
with their content sites?
.return
.skip line
.insert bullet
Classify the key sites you used in resource space,
behaviour space, and in between. Is there any
difference in the relevance or quality of information
from the various types of sites?
.return
.skip line
.indent 0
.hanging indent 0
```

Figure 7.2 Marked-up typescript.

Formative exercise

Revisit the major formative exercises from Chapters 3 and 5, performing them on the Web.

1. How do you combine the search engines and hypertext links to find the information you are seeking?

2. What use did you make, or could you make, of universal or limited-domain portals?

3. How do the portals you use manage their relationships with their content sites?

4. Classify the key sites you used in resource space, behaviour space, and in between. Is there any difference in the relevance or quality of information from the various types of sites.

Figure 7.3 Text rendered differently.

rendering changes could be made once for each element type rather than for each occurrence of each element.

In fact, there are many such semantic markup languages. Most wordprocessors have them (often called *styles*), but the most prominent of them is the *Hypertext Markup Language* (HTML). HTML is based on an earlier standard markup language, SGML (Standard Generalized Markup Language). HTML knows about a few tens of text elements, which are delimited with a start and end mark (start looks like <element> and end like </element>). HTML also has a collection of command markups which can select fonts, etc. The example of Figure 7.2 marked up with HTML would appear as in Figure 7.4. There are three kinds of text elements, a header, a paragraph, and an unordered list, marked respectively by <H3> ... </H3> (the third of six levels of header), <P> ... </P>, and <UL> ... </UL>. The element *unordered list* consists of several *list elements* marked by <LI> ... </LI> within the *unordered list* markup. If you look at a selection of text in a Web editor using *source* mode, you will see a fuller range of marks.

Associated with each text element there is also a set of rendering instructions in a command markup language which can be customized in each browser to make best use of the hardware available and to reflect the user's personal preferences.

```
<H3>Formative exercise</H3>
<P>Revisit the major formative exercises from Chapters 3 and 5, performing them on the
Web.</P> <UL>
<LI>How do you combine the search engines and hypertext links to find the information you
are seeking? What use did you make, or could
you make, of universal or limited-domain portals?
<LI>How do the portals you use manage their relationships with their content sites?
<LI>Classify the key sites you used in resource space, behaviour space, and in between. Is there
any difference in the relevance or quality of information from the various types of sites? </UL>
```

Figure 7.4 Marked up with HTML.

HTML is a remarkable invention. You can create a page using one of dozens of Web editors, upload it to a server running one of several software platforms and expect it to be properly displayed on each of hundreds of millions of browsers from many different sources. None of the software vendors or purchasers need communicate directly with any other.

HTML succeeds because it is a standard. There is a group of people called the World Wide Web Consortium (W3C) which debate and agree on changes to the definition of HTML along with a number of related standards. Being a standard is not enough, though. HTML has also been used in the specification of all the different pieces of software needed to make the Web work. HTML works because its definition is agreed upon by a large body of people and also the definition has been adopted by a wide range of software vendors.

Suppose, however, you invent a new text element, say a spreadsheet. In principle this is easy. Simply define the spreadsheet specification language, say by simplifying the specification language from a widely used spreadsheet, and specify that the rendering is to be done by a spreadsheet program nominated by the user as a browser plugin.

How to get it adopted though? First you must convince the W3C that the markup is a good idea and is well specified. Then the W3C publishes the changed standard, and you must wait for the vendors to change their software to reflect the changed definition. Finally, you must wait for enough users to download the revised browsers to provide a sufficient set of potential users for the material you intend to publish using the new markup. The whole process can take several years. The widespread agreement giving HTML its strength also makes it very slow to change.

What to do? The W3C has recognized the problem and has agreed on a possible solution – the meta markup language XML.

XML

A markup language consists of a set of markup elements and a facility for specifying rendering instructions. HTML works because the markup elements and rendering instructions are hardcoded into all the different software platforms enabling the Web.

XML is a facility for defining markup elements and also for defining their renderings. This more abstract facility is called a meta-markup language – a language for defining markup languages. XML was adopted in 1998 by the W3C and is being implemented by major vendors. When the new products purchased or downloaded by users are installed, what will be hardcoded will not be the markup and rendering, but will be the commands for creating a markup and its rendering.

The elements of basic XML include a method for defining a type of document, and then the markup elements defining the permitted structure of that document type. For example, suppose we want to define a markup language to describe papers submitted to conferences, which look like the sample in Figure 7.5.

This type of document has a structure illustrated in Figure 7.6. There are two main parts, a header and a body. All of the header and part of the body are displayed in Figure 7.5. The header consists of five main parts, two of which, authors and keywords, are themselves structured. The body consists of a series of sections followed by a set of references. Each section consists of a title and a body, which can contain subsections.

Paper to be presented at International Conference on Formal Ontology in Information Systems (FOIS'98) Trento, Italy, 6–8 June, 1998.

Completeness and Quality of an Ontology for an Information System

Robert M. Colomb
Department of Computer Science and Electrical Engineering
colomb@it.uq.edu.au

Ron Weber
Department of Commerce
The University of Queensland
QLD 4072 Australia
weber@commerce.uq.edu.au

Abstract

We examine the problems of completeness and quality in design of information systems.... We can build and use information systems confident that they will be valid under changes in the understanding of meaning and also changes in the understanding of the metaphysics underlying physical and social reality.

Keywords: the ontology of information and information processing, top-level ontological taxonomies, foundations.

1. Introduction

Ever since the introduction of the entity-relationship model [1], the construction of specialised ontologies (often called conceptual models) has been an essential part of the information systems design process. Historically, these ontologies have been partial. They have not provided a model of all phenomena of interest in the domain of discourse.

Figure 7.5 A complex document.

An XML markup language representing the structure in Figure 7.6 is shown in Figure 7.7. The language is defined in what is called a "document type declaration" (DTD). The DTD is constructed from two main commands. One defines a document type (<!DOCTYPE ... >) and the other markup elements (<!ELEMENT ... >).

In Figure 7.7, the document type is paper, whose definition includes all the elements enclosed by [...]. The principal element is also called paper, which consists of two parts enclosed in (...), header and body. There must be one of each, and header must precede body. The latter two elements are then defined, and so on. The definitions bottom out with the built-in element #PCDATA, which means any character text.

There are several other built-in symbols in XML shown in Figure 7.7. Following e-mail is the character "?", which signifies that an e-mail address is optional, but if

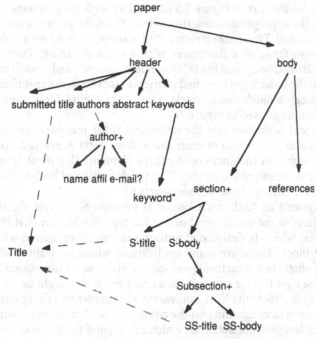

Figure 7.6 Structure of document in Figure 7.5.

present there can be only one. Following author is the character "+", which signifies that there can be many authors but at least one. Following keyword is the character "*", which signifies that there may be many keywords, or one, or none.

```
<!DOCTYPE paper [
<!ELEMENT paper (header body>
<!ELEMENT header (submitted title authors abstract keywords>
<!ELEMENT submitted (#PCDATA)>
<!ELEMENT title (#PCDATA)>
<!ELEMENT authors (author+)>
<!ELEMENT author (name affil e-mail?)>
<!ELEMENT name (#PCDATA)>
<!ELEMENT affil (#PCDATA)>
<!ELEMENT e-mail (#PCDATA)>
<!ELEMENT abstract (#PCDATA)>
<!ELEMENT keywords (keyword*)>
<!ELEMENT keyword (#PCDATA)>

...
]>
```

Figure 7.7 Markup language representing structure of Figure 7.6.

Figure 7.8 shows the text of Figure 7.5 marked up with the language defined in Figure 7.7. With the appropriate rendering instructions, the presentation in Figure 7.5 could be re-created. The markup elements are shown in bold for emphasis.

The example so far is of a document with a rigid structure. There must be elements of header and body, and the DTD specifies that the header must come first. Similarly, within the header the specified elements must be in the specified order. It is possible to specify a much looser markup language. Figure 7.9 is a specification of the markup language used in Figure 7.3.

The first element body says that the document is the text between <body and <\body>, and that there are zero or more marked-up parts. A marked-up part can consist of any number of instances of the three elements U, p or ul, in any order (unordered alternatives signified by the "I" in the definition of body). The element ul is structured, consisting of one or more parts li.

With the fragment of XML described, it is possible to control the structural positions of objects in the document, but not the type. We can say that there must be a block of text which is designated author, but cannot constrain what text is possible in that block. There are many applications where we want to allow only certain types of object in a structural position. An element <illustration> might be constrained to be a gif file, for example, or a cross-reference might be constrained to be a reference identifier (the XML equivalent of a pointer to a designated part of the text). These sorts of constraints on the content of an element are specified with the !ATTLIST declaration, the details of which are beyond the scope of this text.

This section has sketched XML sufficiently to see how it represents structure. The language has many more features, and this brief presentation is not intended to be sufficient for someone wanting to develop XML products – there are many excellent more specialized resources for this purpose. From the perspective of this text, however, we now have a method of representing structured objects and transporting them around the Internet.

<paper><header><submitted>Paper to be presented at International Conference on Formal Ontology in Information Systems (FOIS'98) Trento, Italy, 6-8 June, 1998. </submitted> <title> Completeness and Quality of an Ontology for an Information System </title> <authors><author><name>Robert M. Colomb</name> <affil> Department of Computer Science and Electrical Engineering </affil> <email> colomb@it.uq.edu.au </e-mail> </author> <author> <name>Ron Weber</name> <affil>Department of Commerce The University of Queensland QLD 4072 Australia </affil> <e-mail> weber@ commerce.uq.edu.au </e-mail> </author></authors> We examine the problems of completeness and quality in design of information systems. ... We can build and use information systems confident that they will be valid under changes in the understanding of meaning and also changes in the understanding of the metaphysics underlying physical and social reality. <keywords> <keyword>the ontology of information and information processing </keyword> <keyword>top-level ontological taxonomies </keyword> <keyword> foundations </keyword> </keywords> </header> ... </paper>

Figure 7.8 Text of Figure 7.5 showing markup with DTD of Figure 7.7.

```
<!DOCTYPE loose-doc [
<!ELEMENT body (h3 1 p 1 ul)* >
<!ELEMENT h3 (#PCDATA)>
<!ELEMENT p (#PCDATA)>
<!ELEMENT ul (li+)>
<!ELEMENT li (#PCDATA)>
...]>
```

Figure 7.9 XML DTD for markup language of Figure 7.3.

Applications of XML

So what is being done with XML?

XML was designed to permit people to develop their own markup languages. What might an idiosyncratic markup language consist of? One possibility is a much more structured markup such as partly defined in Figure 7.7. Another is a new element added to HTML, say a footnote rendered in a popup window or a new data type handled by a special plugin, say a spreadsheet. Once there are sufficient XML-compliant browser instances in use, this use of XML might replace some of the use of Java.

There are some costs involved with idiosyncratic markup languages, though. A richly structured markup language must be defined in the first place, which is a significant task. When a document marked-up with this language is accessed, not only the document must be transmitted, but also the DTD and the style sheet rendering instructions. The user's browser must parse the DTD, then use the parsed DTD to parse the document; and also must parse the rendering instructions and use them to render the document. All this requires transmission of much more data and requires much more processing and processor resources than HTML.

A new data type has a much smaller overhead, but the rendering instructions must take into account the great variety of users' rendering devices in order that the new element's rendering will work for all users. Finally, to handle a new data type the browser must have the plugin, which must be transmitted along with the document.

There are people who have made extensive investment in highly structured markup languages. Most publishers use computerized typesetting, and many use markup languages based on SGML or XML to obtain a uniform house style. A large publisher might publish hundreds of journals in a variety of fields. Each journal might publish 100 articles per year, so a ten-year archive for that publisher might contain hundreds of thousands of articles, all marked up with the same DTD.

It is convenient to store these articles in a database. There are specialized database managers supporting SGML or XML and there are more general object-oriented or object-relational databases. Structured extensions to the standard relational databases can store the structured articles. If a user requests an article from the archive, it is retrieved from the database as an XML document, then rendered by the server into HTML, and sent to the user. Figure 7.10 shows the architecture of such a service.

In this way, the XML DTD is used to specify the database schemas with which the archive is stored as a collection of structured documents, but the user does not need

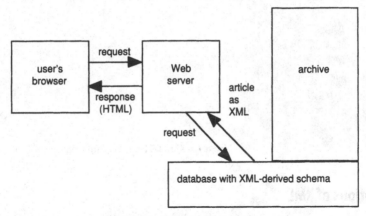

Figure 7.10 An XML-structured archive.

anything more than a standard HTML browser to read an article. Note that the XML document is rendered into HTML. Remember that the browsers already have methods for rendering HTML documents, so the XML rendering can make use of them.

Building on this, if the archive is stored with its structure represented, then the users might want to query the database using the structure. If the users know the DTD, which can be obtained from the publisher's site, they might want to find articles containing particular terms in section titles or which cite particular references. There are a number of query languages existing or under development which combine the text capabilities of the ISAR systems of Chapter 3 with the structure representation capability of SQL and its relatives. To say much more about these languages is beyond the scope of this text. However, a user can make a query taking into account the structure of the documents in the archive, and the documents returned can be rendered into HTML for transmission to the user's browser.

Rendering a structured document into HTML can be much more than varying headings, fonts and margins. It is quite possible to render the section and subsection heads as tables of contents with hypertext links to the corresponding section contents, for example, or to make active cross-references links. The XML marked-up text contains sufficient information to enable this.

Therefore, a complex document can be rendered as a hypertext residing on the user's computer. Also, a complex hypertext such as a Web site can be packaged and transmitted as an XML document. Exchange of complex documents becomes much easier.

Probably the most active development of XML at the time of writing is implementing EDI – Electronic Document Interchange – to enable electronic commerce. Business is carried on by the exchange of standard documents among the parties. Company A might want to buy 1000 widgets in January 2001. Companies B and C supply widgets. Company A will start the process by sending a *request for quotation* to B and C. Both B and C respond with *quotation* documents. A chooses one of the quotes, say that of B, and issues a *purchase order* to B. B ships the widgets on 10 January, together with an *invoice*. Later, A will request their bank to send a *funds transfer* to B's bank.

Each of the standard document types – request for quotation, quotation, purchase order, invoice and funds transfer – has a standard interpretation supported by the commercial law apparatus of the various jurisdictions in which the companies operate, since they are all forms of contract. Each document has a standard set of items it can contain. For example, an invoice might be required to contain date of shipment, selling company, purchasing company, an identifier, identifiers and quantities of the products shipped, and the total value of the invoice. An invoice might optionally contain the identifier of the purchaser's request for quotation, the supplier's quotation and the purchaser's purchase order to which the invoice responds. It might also optionally contain a delivery address. In practice there are hundreds of types of items which can appear in an invoice document. The other documents are similar.

EDI is a standardization of these business messages and their possible contents. Using EDI, businesses can exchange documents directly from information system to information system without having to rely on human data entry. Since these documents are highly structured, it is natural to use XML as a definition and transmission medium.

As shown in Figure 7.11, the standards body defines the message types and possible contents, then represents the standard as a DTD which is stored in a repository. Sellers and purchasers download the DTD, encode their business transactions as XML documents and transfer them using Internet protocols.

A much-simplified partial DTD of an EDI invoice together with an example of a marked-up instance is shown in Figure 7.12. In practice, XML EDI messages are more complex and more constrained, but the figure gives the flavour.

Therefore a group of people or organizations can conduct business over the Internet, exchanging precisely specified structured messages. They only have to do business according to the practices contained in the standard, and to download the DTD supporting the message types. Furthermore, the collection of messages can be stored and queried using standard structured or ISAR database methods.

EDI technology is not limited to electronic commerce. Any application where structured data is exchanged can be implemented using similar methods. As XML technology develops, we may see applications presently using HTML forms migrate to XML, or at least the more complex ones.

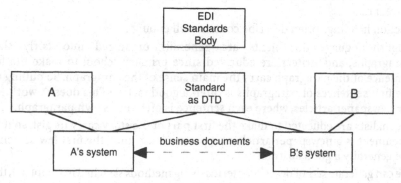

Figure 7.11 Sketch of EDI implementation.

```
<!DOCTYPE edi-invoice [
<!ELEMENT invoice (buyer, seller, date, product, amount)
<!ELEMENT buyer (#PCDATA)>
<!ELEMENT buyer-purchase-order? (#PCDATA)>
<!ELEMENT seller (#PCDATA)>
<!ELEMENT date (#PCDATA)>
<!ELEMENT product (#PCDATA)>
<!ELEMENT amount (#PCDATA)>
...]>
<invoice> <buyer> Company A <\buyer> <seller> Company B
<\seller> <date> 2000.04.05 <\date> <product> Widget Model 42
<\product> <amount> USD 548.00 <\amount> <\invoice>
```

Figure 7.12 XML EDI Invoice.

Automatic abstracting

There are many applications, generally when we are searching for information, where we want to see small abstracts of possibly large documents. The abstract gives us a quick indication whether it is worth downloading the whole document and investing the time needed to read it. The abstract may be included in the catalogue entry. Many search engines, for example Google, display their search results with brief abstracts.

Abstracts are valuable, but are expensive to produce. It would be nice to produce abstracts automatically from the documents. To produce high-quality abstracts automatically, though, would require that the programs attain something approaching a human's understanding of the text, which is well beyond present-day technology. But often the abstract can be a bit funny, so long as it is readable. Automatic systems exist which can do a pretty good job.

Automatic abstracting programs use elaborations of the sorts of query and text processing technologies described in this textbook. They use essentially a collection of rules of thumb, such as:

- If there is a section with a heading titled "Executive Summary" or "Abstract", extract it.
- Section headings often describe content well enough.
- English-language documents are generally organized into fairly short paragraphs, and writers are educated since primary school to make the first sentence of the paragraph carry the main sense of the paragraph. So pulling out the first sentences of paragraphs is often a good tactic. (This doesn't work well for newspaper articles, where each sentence is often in its own paragraph.)
- Journalists are educated to make the first part of the story carry its gist, so if the document is a newspaper article or a press release, then the first few sentences are generally a good abstract.
- We can generate sets of descriptor terms using methods sketched in Chapter 4, then extract sentences which contain a disproportionate number of descriptor terms.

All of these tactics suffer from the usual problems of precision and recall. If the documents are marked up with XML and if the analysis program understands the semantics of the DTD, then precision and recall can be improved. In particular, we can reliably identify executive summaries, abstracts and section heads at different levels. If the document has an index, we can get a reliable list of important terms. If sentences containing citations can be identified, they probably contain good content and can be extracted.

Even if the DTD is not some established standard, if the collection has a large number of documents marked up with the same DTD, it may repay the investment of an analyst to understand the markup language and to tell the analysis program what to look for.

XML may therefore make the problem of automatically building abstracts a little easier.

Information space with XML objects

This chapter has looked at a method of representing the structure of complex documents of various sorts. This opens more possibilities in the organization of an information space. We of course have all the facilities of the global information retrieval and hypertext systems described in previous chapters.

In addition, the documents become more than point objects. When the larger system leads a user to a structured document, a more specific structure becomes visible, and the browser/ information retrieval system can allow the user to navigate within the document.

Furthermore, with XML representation of structure it is possible for the DTDs describing the structure of a class of documents to be made globally visible. In this way the information retrieval and hypertext navigation tools in the larger system can penetrate into documents of that class. This is especially useful if documents of the globally described class can be present on many sites.

Representation of structure with XML makes the information space deeper, more flexible, and more informative.

Key concepts

The structure of complex documents are described using **markup languages** such as SGML or **XML** These languages allow the user to define a set of **markup elements** using a **document Type declaration.**

Further reading

There are many books, articles and Web sites with information about XML and its applications. Good basic reference books include Goldfarb and Prescod (1998) and Megginson (1998). Automatic abstracting is discussed by Salton (1989).

Formative exercise

Find an application of XML, perhaps by searching on the Web.

- What does the application entail?
- Who develops and maintains the DTD? Where is it stored?
- Are there multiple parties involved in the application? If so, how do the parties obtain the DTD?
- Does the application support queries? If so, how do the databases get accumulated?
- What query language is used?

Tutorial exercises

Sketch a markup language useful for e-mail.

1. Turn this into a rigorous (specifies not only what elements are permitted, but specifies a structure with possibly optional components) markup language (note that e-mail messages may quote other messages).
2. Imagine the e-mail is to be sent to a Web site for automatic inclusion into some sort of nice structure (for concreteness, think of a mailing list designed to update a Web site devoted to a course using this textbook). What are the implications for the markup language?
3. What help does the markup language designed so far give for searching? How could it be improved?
4. How would you go about summarizing the messages sent to the Web site? Would the markup language help?

Open discussion question

Generalized markup is intended to describe the content of a document so that it can be rendered successfully in a wide variety of media, changing only the instructions for rendering the named text elements. Is it feasible that this variety could include voice generation? (Voice generation has been in practical use for some years, and is improving all the time.) Suggest ways of rendering in voice the XML or HTML examples which have been presented. Does it work? What goes wrong? Can these problems be fixed?

Further, some textual elements seem to require a definite spatial arrangement – think of a table with several columns and several rows. Could this be marked up in a medium-independent way? For voice?

8 Controlled vocabulary

Up until now, we have been looking at the information structures defining an information space from the point of view of someone seeking some information. In this chapter, we will be looking at these structures themselves, and seeing how they provide the geometry of an information space, and how they can be used to gain an overview of the space. These information structures are generally "controlled vocabularies".

Why controlled vocabulary?

A controlled vocabulary is a collection of terms from which descriptors are drawn in the construction of surrogates. In general, a document will be described by several terms from the controlled vocabulary. There is an extremely important special case of controlled vocabulary, the classification system, in which each document is described by exactly one term from the vocabulary.

This chapter looks at the general problem of the design of a controlled vocabulary, and classification systems most particularly. Subsequent chapters will examine how the need for visualization of the information space affects the classification system design; the extent to which classification systems are designed, not discovered; a detailed theory of the design of classification systems; then of subject terms and thesauri; and finally how these classification systems can be used to visualize the contents of an information space.

We have already encountered the issue of a controlled vocabulary in the discussion in Chapter 4 about assigning descriptors to document surrogates. There, the emphasis was on describing the document so it could be retrieved, but some overall system issues were considered, such as specificity and exhaustivity, computer aids to generating term sets, and the use of term sets to get an overall view of a collection of documents using content analysis. In this chapter, we focus on the system issues.

The information retrieval view of controlled vocabulary is to provide a collection of standard search terms, while the information space view is that they provide semantic organization of an information space. Consider, for example, the simple system shown in Figure 8.1, a menu from a noodle restaurant.

First of all, Figure 8.1 shows a controlled vocabulary, subdivided into four lists. Every dish served in the restaurant can be described by a combination of terms from the lists. More specifically, each dish in the restaurant is characterized either

Menu

- Choose kind of noodle
  - Egg noodle (thin)
  - Shanghai noodle (Northern China nodle)
  - Hokkien noodle (smooth oil base egg noodle)
  - Rice Noodle (Thailand and Vietnam noodle)
- Choose your favourite style of cooking
  - Thai (hot and spicy)
  - Japan (simple but tasty)
  - Chinese (complex and delicious)
- Select your favourite main ingredient
  - BBQ pork
  - Roast duck chicken beef prawn
  - Vegetarian seafood combination
- Noodle soup
  - Singapore Laksan
  - Wonton noodle soup
  - Meat ball noodle dish

Figure 8.1 Noodle restaurant menu.

by a selection of exactly one term from the first three lists, or exactly one term from the last list.

A diner choosing a dish uses this controlled vocabulary for what amounts to information retrieval purposes – they either make a sequence of choices from the first three lists or a single choice from the last. The vocabulary also gives an overview of the space, in at least two ways. First, a customer can use it to decide whether to come to the restaurant at all; by seeing that interesting choices are available without making a commitment to a specific choice. Second, if the restaurant keeps track of the meals served, the vocabulary allows a tabulation of the number of meals associated with each choice or combination of choices. Such a tabulation is useful in planning supplies, or for suggesting changes. If, for example, some choices are made very infrequently, their removal can be considered, while if other choices are made very frequently, additional variations may be proposed – say if Singapore Laksan becomes very popular, then additional types of Laksan could be added to the menu.

It is also possible to view the vocabulary of Figure 8.1 as a classification system. Certainly each dish from the last list is characterized by exactly one term. Each other dish is characterized by exactly three terms, one from each of the first three lists. It is possible to construct compound terms by, say, concatenating the three terms in a predefined sequence. The resulting concatenated terms ranging from <egg noodle : Thai : BBQ pork> through <rice noodle: Chinese: combination>

form a classification system which, with the addition of the terms from the last list, is exhaustive.

What do classification systems look like?

Classification systems are particularly important, so that much of the discussion in this and the following chapters is concerned with this special case. Plato and Aristotle invented classification in the 4th century BC as a formal method of understanding things. Possibly the first classification system published was in Plato's *Phaedrus*. The system is shown in Figure 8.2.

In the dialogue, Socrates, having earlier excoriated jealous and self-centred love as evil, is attempting to explain a good type of love. He first convinces Phaedrus that love is a sort of madness, in that it makes people behave irrationally. He then divides madness into madness due to human infirmity and madness due to divine inspiration. Then madness due to divine inspiration is further divided into prophetic trances, initiatory mystical experiences, poetic inspiration by the Muses, and a final class, the erotic, which is what we know as Platonic love.

The method Plato advocates is that when one wants to understand something, one first generalizes by finding a broader class into which the particular thing fits, then dividing that general class into species "at the joints" – that is, by well-defined characteristics so that agreement can be obtained on the subdivision.

For example, consider the picture in Figure 8.3. How are we to understand what that picture represents? By Plato's method, we would first introduce a general class, say *animal*. Everyone would agree that Rosie is an animal. We might then divide the class *animal* into *wild* and *domesticated* - Rosie is clearly a domesticated animal. This new class might be divided into *farm animal* and *pet*. Rosie is a pet. The class *pet* might be further subdivided into *dog, cat, bird, fish*, and *other*. Rosie is a pet dog, as distinguished from a farm working dog or a wild dog. Each of these divisions shows either what Rosie is, or what she could be but isn't, thereby increasing our understanding of the concept Rosie.

Notice that the subdivision of the non-pet branches is not carried out. This is often the case in Plato. Aristotle, on the other hand, used this method of genera and species to make extensive classifications of all sorts of phenomena, subdividing each branch down to a level comparable with each other. In this sort of

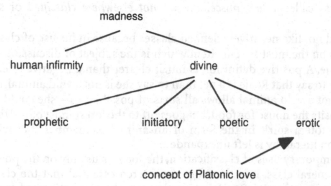

Figure 8.2 Classification system from the *Phaedrus* explaining Plantonic love.

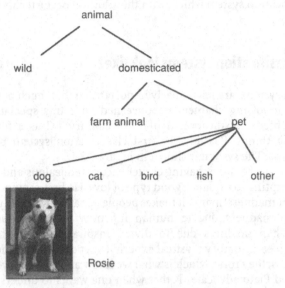

Figure 8.3 Understanding Rosie.

classification, we can concatenate the terms in each sequence of subdivision to obtain a single descriptor for the phenomenon. From Figure 8.2 we obtain <madness : divine : prophetic> through <madness : divine: erotic>. From the discussion of Figure 8.3 we obtain <animal : domestic : pet: dog> through <animal domestic : pet : other>.

The classification system of Figure 8.1 can also be seen in this light. The most general classification is *noodle dish*, which is subdivided into *dry* and *soup*. *Soup* has three species, while dry is further subdivided by *kind of noodle*, then *style of cooking*, then *main ingredient*.

A peculiarity of the Rosie classification system is worth noting. The last subdivision, *pet*, has four definite subclasses: *dog, cat, fish, bird* and one indefinite class, *other*. This last covers any pets which are not one of the four definite subclasses. This sort of class is defined negatively, and is warned against by Plato in *The Sophist*, for example, as contributing less to understanding than positively defined classes. However, nearly every system of classification has a negatively defined class called *other, miscellaneous, not elsewhere classified* or something similar.

Plato did not like negatively defined classes because in his use of classification, the focus is on the most specific item which is the subject of discussion – Platonic love or Rosie. A positive definition is much clearer than a negative one. It tells us much more to say that Rosie is a pet than to say she is not a wild animal, because to say she is not a wild animal allows all sorts of possibilities – she could be a tame wolf that visits the house for food, but goes off to the forest most of the time, or she could be a totem spirit in the form of an animal. Remember that most of the classification hierarchy is left unexpanded.

In contemporary uses of classification, the focus is usually on the population of the most general class. It is therefore considered essential that the classification system be exhaustive. It often happens that the characteristic used to create the more specific classes is not well defined for all the members of the population, so the negatively defined class is used for those things that are "neither fish nor fowl". One might classify the people attending a university lecture by the course in which they are enrolled, but there might be a few people not enrolled in any course – say, colleagues of the lecturer or friends of students who happened to be interested in the day's topic.

Note that there is a second peculiarity in the Rosie system of classification. Within the class *pet*, we have *dog, cat, bird* and *fish*, as well as the negatively defined class *other*. The first two classes, *dog* and *cat*, are much more definite than the others. To be consistent, we should have classes *canary, parakeet, falcon, goldfish, angelfish, piranha*, and so on. However, if we did this, we would have very many classes, and only the classes *dog* and *cat* would have very many members. Relatively few people keep canaries or piranha. So the system of classes would be unbalanced. Most pets are in the first two, and only a few in each of the many other classes. To overcome this imbalance, it is common to use less definite superclasses in place of a number of sparsely populated more definite classes – hence the class *bird* in place of *canary, parakeet, falcon* and the others; and the class *fish* in place of *goldfish, angelfish, piranha* and the others. The final, negatively defined, class *other* is needed because a few people keep pet ferrets, or pigs, or snakes, or tarantulas, and there is no convenient superclass that includes all of them.

Note that neither the Platonic *love* nor the *noodle* systems have explicitly negatively defined classes. In the former case, the system is not exhaustive. For example, one would probably class the sort of insane courage that often wins bravery awards as a species of *divine madness*, but it is not one of the four named. In the latter case, the system is built from the classes. The menu is established before the dishes are prepared, and by design no dish is prepared which is not on the menu.

Sometimes negatively defined classes are not obvious. The test comes when a new document is to be catalogued which does not fit into the classification system. What instruction is given to the cataloguer? If to assign the document to a default class, then that class is negatively defined. If, for example, the noodle restaurant introduced fried rice, that dish would probably be assigned to the *noodle soup* class – the class is essentially all the dishes the restaurant serves which do not fall within the *noodles* ? *sauce* ? *main ingredient* classification system.

The Standard Industrial Classification (SIC) system used in economics for industries (extracted in Appendix A) has negatively defined classes. Division 99 in major group J is *non-classifiable establishments*. A new type of firm is assigned to Division 99 if it does not fit into any other class. Industry group 7389 *Business Services, Not Elsewhere Classified*, in major group 73 of Division 1 is also negatively defined. Notice further that major group 73, *Business Services* is itself negatively defined – its description (Appendix A) includes:

> This major group includes establishments primarily engaged in rendering services, not elsewhere classified, to business establishments …

Another possibility is that if the document does not fit the classification system, it is rejected – in this way the classification system is used to express the limits of the domain of the document collection. The classification system is not exhaustive with respect to the set of documents which come under consideration, but is used to choose which documents will be retained in the collection. Finally, the instruction might be to attach the document to whichever classification seems to the cataloguer at the time to be nearest. In this case, the classification system is being undermined by the principle of uncertainty – if it is not clear which class the document belongs in, then two cataloguers will probably assign it differently, and a user might look for it somewhere else entirely.

This last often occurs with library systems in rapidly developing areas – the library is committed to holding documents in the areas taught by, say, a biological sciences faculty, but the classification system available is externally developed by, say, the US Library of Congress, and is between its periodic reviews and reorganizations.

How are classification systems organized?

This section examines some high level issues in the organization of classification systems. More detail is presented in Chapter 9, and especially in Chapter 11.

The method of designing classification systems by genera and species is called, in information science, theoretical analysis. The classification system is constructed from an a priori examination of the domains covered by the collection of documents, often before the collection is established. The most general class is the field of the collection, the "joints", are identified by consideration of the most salient characteristics of the material contained in the documents, the subclasses named and the process repeated for each subclass.

As alluded to in Chapter 4, how far this process goes depends on the size of the collection. The level of specificity desired is a design parameter, so that the desired size of the subcollection of documents associated with each most specific class is established before the classification system begins to be organized.

The Standard Industrial Classification scheme in Appendix A is designed by theoretical analysis. Notice that Major Group 08 *Forestry* is subdivided only one level, while Major Group 73 *Business Services* is subdivided two levels. There is certainly a variety of firms in the class *Forest Services*, but not enough firms to be worth further subdivision. The *Encyclopedia Brittanica Propaedia* of Appendix C is also designed by theoretical analysis and has great differences in depth. It is divided to five levels via Section 10.36 B, *The nature of anthropology*, but seven levels in the preceding subsection A, *History of the social sciences*.

This illustrates that when a theoretical analysis is applied to a collection of documents, stopping at a given level of specificity can result in some high-level classes having more levels of subdivision than others, sometimes many more. This imbalance can be a problem for a number of reasons, including:

- A user interface based on a theoretical analysis can have a highly unbalanced menu structure, where some functions are buried very deeply in the menu hierarchy, while others are quite close to the top. A menu version of the noodle restaurant would look something like Figure 8.4, with the Japan egg noodle choices fully expanded. It takes two clicks to choose wonton noodle soup and four to choose vegetarian Japan egg noodle.

- An unbalanced class system can result in unused space in the code system used to represent the classes. The Standard Industrial Classification is represented by a four-digit code. The digits devoted to Division A are much more sparsely filled than those devoted to Division I. (These code systems are discussed more fully in Chapter 11.)

- Wet noodle
  - Singapore Laksan
  - Wonton noodle soup
  - Meat ball noodle dish
- Dry noodle
  - Egg noodle
    - Thai
    - Japan
      - BBQ pork
      - Roast duck
      - Chicken
      - Beef
      - Prawn
      - Vegetarian seafood combination
    - Chinese
  - Shanghai noodle
  - Hokkien noodle
  - Rice noodle

Figure 8.4 Hierarchical choice version of noodle restaurant menu.

An alternative way to create a classification system is to begin with the collection of documents and a desired level of specificity. The first step is to group the documents into small subsets at about the desired specificity, and to name each of those subsets. This gives the most specific set of classes. The next step is to group the subsets into somewhat larger subsets and name those subsets, giving the next most general set of classes. The process proceeds until the number of highest-level classes is small enough. This basis of obtaining a classification system is called *literary warrant*. (A *warrant* is a basis for making a judgment, and the warrant is called *literary* because the judgment is based on consideration of the actual collection of documents to be classified.)

The organization of a book into chapters, sections and subsections tends to be done using literary warrant. This book is itself a good example. Each most specific section is a discussion of a very specific topic. These subsections are grouped into chapters which are mostly about the same number of pages. If the material were organized according to a theoretical analysis, the most general levels of the table of contents would look something like Figure 8.5, whereas in the actual case, all the chapters are at the same most general level.

Literary warrant-based classification systems tend to have less intuitively understandable terms in them than do theoretical analysis-based systems. Appendix B shows the top level of the classification system used in the search engine Yahoo!, together with an expansion of the *WWW* subclass of the *Computers and Internet* class. Both collections of terms look quite arbitrary. (Yahoo! is actually

- Overview
- Details
  - ISAR systems
    - Query languages
    - Cataloguing
  - Hypertext systems
    - Hypertext
    - World-Wide Web
  - Markup languages
  - Controlled vocabulary
    - Classification
      - Theory of
      - Visualization aspects
      - Philosophical aspects
      - Techniques
      - Use in visualization
    - Subject indexes
- Other issues

Figure 8.5 A theoretical analysis version of the table of contents.

built using a mixture of theoretical analysis and literary warrant techniques, as is shown by the expansion of the *College and University* subclass of the *Education* class, which is a logical division of aspects of tertiary education.)

One would tend to use a theoretical analysis approach when planning for a collection (there is by assumption no actual collection to analyse), or when focusing a collection on a specific purpose. In these cases, the emphasis is on the use of the collection. The material to be gathered will be chosen to fill the classes in the system. Material in an existing collection which does not fit into the designed classification system may be discarded.

On the other hand, where the collection is primary, one would tend to use a literary warrant approach. Here the collection already exists, or is perhaps the archive from some ongoing activity (a newspaper, say). The issue here tends to be how to present the collection to potential users, who may not be well defined beforehand.

Successful classification requires either a well-defined user requirement or a well-defined collection. The World Wide Web as a whole is almost impossible to classify effectively, at least for a general audience.

In summary, we have two methods of designing classification systems: theoretical analysis which starts from the domain and works from the most general to the most specific; and literary warrant, which starts from the population and works from the most specific to the most general. Most systems in practice use a combination of the two methods.

Classification systems are often hierarchical

All of the examples of classification systems we have seen are hierarchical – they consist of a series of progressively more specific groups of terms, organized so that groups of more specific terms are associated with single more general terms. Some classification systems are not hierarchical.

- Male versus female
- Animal, vegetable or mineral
- Large, small, average
- Native-born citizen, naturalized citizen, foreigner.

These non-hierarchical systems are very small. Even so, the last example is probably more usefully expressed as a hierarchy.

- Citizen
  - Native-born
  - Naturalized
- Foreigner

Hierarchy is nearly universal in classification systems because a fundamental use of classification is to gain an overview understanding of an information space – classification systems are designed to be understood by humans. Humans cannot keep a very large number of distinctions in mind at any one time. The smallest number of distinctions is two, the dichotomy. Most informal classifications encountered in everyday life are dichotomies. As we have seen above, even systems as small as three are often constructed from a hierarchy of dichotomies. Most formal systems have fewer than ten distinctions at any level – the SIC of Appendix A is typical of medium-sized systems, and the Dewey Decimal System used in libraries (described in Chapter 11) is a very large system with no more than ten distinctions at any level. A few systems have more distinctions. Yahoo! in Appendix B has up to 50 or 60 classifications at the more specific levels, and the Library of Congress system described in Chapter 11 has about 50 most general divisions (the Schedules – see Appendix E).

Systems with more than 100 distinctions at one level are extremely rare. Consider the United Nations, which has a few less than 200 member nations. They are almost always grouped in various ways for various purposes – by level of economic development, by race or religion, by continent or major region, by political system – into a small number of more general classes.

Classification systems can be very large. The major library systems have hundreds of thousands of most specific classes, as does the medical classification system SNOMED, discussed in Chapter 11. The Linnaean system, whose most specific elements are genus and species, organizes millions of biological species.

As we saw in Chapter 4, the number of most specific classifiers is determined by the desired degree of specificity and by the size of the collection, either actual or planned, a result of the principle of variety. The depth of the hierarchy is determined by the size of each step, which has a maximum due to the limitations of human understanding. The remaining design decision is whether to opt for a small number of divisions of each more general term (a *narrow hierarchy*), leading to a *deep hierarchy*; or for a large number of divisions (a *wide hierarchy*), leading to a *shallow hierarchy*.

A narrow hierarchy minimizes the impact of the principle of uncertainty. If the task is to classify an object into only two classes, there is much less opportunity for grey areas than if the task is to classify into twenty (this point is discussed in detail in Chapter 10). The behaviour of the indexer and the person making a query thus more reliably matches, increasing both precision and recall. The disadvantage is that the resulting hierarchy can be very deep – it takes ten binary choices to identify one of one thousand most specific classes, and twenty to identify one of one million.

The *Encyclopedia Brittanica Propaedia*, a fragment of which is shown in Appendix C, is fairly narrow and hence quite deep – Section 10/36.A.3.d is six levels deep and is subdivided, giving a depth of seven levels. This is very deep considering that the system classifies fewer than 100,000 articles. It works because it is presented on large pages in small type, so that moderately large portions can be seen at one time (the breakdown of Section 10/36 shown in the Appendix occupies less than one page).

Shallow hierarchies with dense menus have become more common as people become more familiar with computerized information sources – the Yahoo! system of Appendix B is a good example. The top level of the hierarchy has 14 choices, each of which displays its most frequently accessed and three or four next most specific choices, so that there are 59 choices available. Even so, the system is so large that the hierarchy has an average depth of about six. The principle of uncertainty is minimized here by a sort of cultural agreement over time as successful matches between indexer and user are reinforced and unsuccessful matches are not.

Basis of hierarchy

Now that we know strategies for designing a classification system and how to choose its size and shape, we turn our attention to the choice of terms to be used in it.

First, note that any term used designates a subset of documents – every document is assigned to exactly one most specific term, and each more specific term is assigned to exactly one term at each more general level. One way to choose terms for the classification system is to reflect this fact in the terms used. That is to say each more specific term designates a subtype of the more general term via a specific value of a general characteristic. The Platonic love example of Figure 8.2 works that way. Human infirmity and divine are two sources of madness, and prophetic, initiatory, poetic and erotic are four modes of expression of divine madness. The terms used in the hierarchical version of the noodle restaurant example of Figure 8.4 designate subtypes, respectively values of wet/dry, type of noodle, style of cooking, and main ingredient. The larger Standard Industrial Classification example of Appendix A is also organized this way. Subtypes are familiar to anyone who understands database conceptual design using, say, the entity-relationship method.

There are other relationships among terms that can be used as the basis for subclassification, however; notably the set/instance relationship and the whole/part relationship. We introduce these two by example.

Consider a collection of documents about computers.

The classification system in Figure 8.6A is the subtype relationship we have already seen, with different values for the characteristic CPU. In Figure 8.6B, the subclass terms are all instances of the class term *famous computers*. There is a

A. Subtype relationship
- Computer system
- Intel CPU
- Motorola CPU
- Alpha CPU
- PowerPC CPU

B. Set/ Instance relationship
- Famous computer systems
- ENIAC
- Colossus
- HAL
- Deep Thought

C. Whole/Part relationship
- Computer system
- CPU
- Terminal
- Printer
- Disk

Figure 8.6 Three types of sub classification.

considerable body of literature about each of these, so that each term identifies the subclass of documents about that computer. In Figure 8.6C, all the subclass terms designate parts of the thing designated by the class term. A product catalogue is often organized in this sort of way – "printer" identifies the subset of products in the catalogue which are printers, etc.

Parts of the Yahoo! system of Appendix B are organized in each of the three ways. *Social Science* and *Reference* are both organized as subtypes. *US States* is eventually resolved into individual states, giving a set/instance type of classification of Web pages. *Colleges and Universities* at the most general level is organized quite well as whole/part.

The *Encyclopedia Brittanica Propaedia* of Appendix C is also organized in each of the three ways. Part Ten, *The Branches of Knowledge,* is subdivided by set/instance – mathematics, philosophy and so on, are members of the set branches of knowledge. Section 10/36, subdivision A, *History of the social sciences,* is subdivided by whole/part relationships down three levels. Subdivisions B, *The nature of anthropology,* and C, *The nature of sociology,* are also whole/part, but much shallower. Part Ten Division III, *Science,* is organized by subtypes (although 10.31 *History and Philosophy of Science* is an anomaly – there is some element of literary warrant in the design of the *Propaedia*).

Of course, it is not necessary to use a homogeneous method of subdivision. It is possible to imagine a collection of documents for which a literary warrant approach would have the classifications shown in Figure 8.7. The Library of Congress system,

- Computer system
  - Computer system
  - Intel CPU
  - Motorola CPU
  - Alpha CPU
  - PowerPC CPU
  - ENIAC
  - Colossus
  - HAL
  - Deep Thought
  - Terminal
  - Printer
  - Disk

Figure 8.7 A hypothetical literary warrant-based classification system.

which we will look at in Chapter 11, is built mainly from literary warrant, and has many classifications with heterogeneous subdivisions.

The advantage of a homogeneous method of subdivision of classes is that it aids comprehension. Suppose one had to classify, using the system of Figure 8.7, an article on the terminal equipment used for Colossus – would it go under *Colossus* or under *Terminal*? Would someone seeking this information look in the same class? The same problem occurs with the expansion of *Computers and Internet: WWW* in Yahoo! (Appendix B). Where would a book on Java go? Or meta-content format implemented in XML?

For this reason, it is probably better to use a mainly theoretical analysis approach to any collection of documents which has a well-defined content, reserving a mainly literary warrant approach to collections whose content is difficult to predict, such as a general library, the contents of a newspaper, or the World Wide Web.

Designing a classification scheme

Designing a classification scheme for a particular application is a special case of the problem of choosing descriptors for a document in a surrogate as discussed in Chapter 4. If we know something about the intended user population, then we can think about classifying according to the viewpoint of the users. By analogy with literary warrant, we can call this approach *user warrant*. An example from Chapter 4 is to classify newspaper articles that have the potential to affect an organization by several degrees of urgency.

It turns out, though, that it is difficult to construct classification systems by user warrant, as distinguished from designing keyword-style subject descriptors. The problem is that in a classification system each object is assigned to just one class. An organization may be able to develop a classification system for its concerns, perhaps around product groups or organizational subdivisions or stakeholders. It is often the case, however, that a particular object is relevant to several of the organization's classifications.

There is also a time dimension to the problem. An organization's concerns change over time as its structure adapts to its environment. In particular, tactical considerations change with time. An article may be of urgent interest because it is relevant to a decision which, once made, is difficult to alter. The day after the decision is made, the article's relevance will change to "of historical value", and a new very similar article may be of no interest whatever. Unless the collection is very transient, relevance may not be a good way to classify (although it may still be a good descriptor).

Classification systems therefore tend to be constructed on the basis of intrinsic properties of the objects in the collection. User warrant is exercised mainly in the choice of objects to put into the collection, and in the descriptors included in the surrogates.

It is often the case that there are multiple classification schemes applying to a collection. The noodle restaurant example of Figure 8.1 can be seen as incorporating three distinct classification schemes: kind of noodle, style of cooking, and main ingredient. Each of these systems is exhaustive, and each dish is a member of a *composite class*. The hierarchical choice version in Figure 8.4 is based on an arbitrary choice of order to apply the schemes.

The classes in the different systems are related by the common occurrence of items. This issue is taken up in Chapter 9, where the different systems are called semantic dimensions, and the concern is the organization of the information space for visualization, as further developed in Chapter 13. It is also taken up in Chapter 11 in connection with design of classification systems for large collections, where it is called the *faceted approach*.

Some collections are of items which have considerable substructure – long documents, complex images, long audio or video clips, complex criminal records, for example. One may wish different fragments of large objects to have different classifications.

For example, proceedings of a conference on the greenhouse effect and energy policy might be classified under *energy policy*. However, individual articles may be mainly about coal, or metallurgy, or transportation, or information technology. It can make sense to separately classify each article in addition to the classification of the proceedings itself. A television current affairs programme may be classed as *current affairs*, while individual segments might be classified as *foreign affairs*, *sport*, or *human interest*.

Sometimes a large document's decomposition into parts is aligned with the hierarchical decomposition of the classification system, so that the classes assigned to the parts are subclasses of the class assigned to the whole document. For example, a sporting yearbook might be classed as *sport*, but its sections classified by the individual sport discussed in each.

Finally, we have the question as to how deep the hierarchy should go – how many classes there should be. This of course depends on what we are trying to achieve. In most cases, the principle of variety applies: we have a collection of objects and an idea as to how many objects we want in the most specific class, so the number of classes depends on the size of the population.

However, sometimes we want the classification system to tell us about the space of possibilities for objects, so we are interested in classes which have no members or in extreme cases in classes which are unusual in that they have members. An example is the Olympic Games. People often use two different systems to classify competitors – by sport and by country. There are a large number of sports, so that the number of competitors in most sports is small. There are also a large number of countries, so that the number of competitors from most countries is also small. If we use both classification systems together, many of the composite classes have no competitors.

Public health disease investigation makes great use of this strategy. People who come down with an unusual disease are assigned to classes where there are composites of many different systems – sex, age, occupation, area resident in, area working in, recency of overseas travel, etc. The investigators hope that the cases will cluster in a small number of composite classes, so that nearly all of the composites are hoped to be empty.

The principle of uncertainty can also affect the number of classes. If the classes are used as query terms, then the error effect of specificity is relevant. From Chapter 4 we recall that for a successful classification of an object, both the indexer and the user must choose the same class – the index puts the object into a class and the user must look there to find it. The more specific the class, the more likely the indexer and user will not agree.

Some systems of classification are less affected by the principle of uncertainty; for example, those intended to present a collection of objects to users. A lecturer in an information science subject may accumulate a collection of newspaper clippings, presenting them to the students classified by the lecture topic to which they are most related. The lecturer expects the students to get to the articles from the lecture topic – indeed the articles are intended to help define the lecture topic. The student is never expected to independently seek a specific article by figuring out which lecture topic it pertains to.

This observation tends to apply generally to classification systems which are used to provide structure to a visualization of an information space, as discussed further in Chapter 13.

Key concepts

A collection of documents with a way of telling which documents are near each other (a topology) or where documents are or might be (a geometry) is called an **information space**. One way to organize an information space is with a set of descriptors called a **controlled vocabulary**. If each document is assigned exactly one descriptor the controlled vocabulary is a **classification system**, if generally more than one a **thesaurus**.

Large controlled vocabularies are usually organized into a hierarchy based on **generalization** (terms have lower specificity) and **specialization** (terms have higher specificity). The **depth** of the hierarchy is the average number of steps from the most general to the most specific descriptor, and the hierarchy is **balanced** to the extent that all paths from most general to most specific are the same length.

Controlled vocabularies are developed in several ways. The method called **literary warrant** is based on analysis of an existing collection of documents so that the most specific descriptors in the hierarchy describe approximately the same number of documents, and the hierarchy is balanced. The method called **theoretical analysis** is based on an a priori principled subdivision of more general terms. A method taking into account the relationship of the documents to potential users is called **user warrant**.

Further reading

Most texts in library science cover the concept of classification, as do basic texts in botany or zoology. It is worth looking at large examples, in particular the *Encyclopedia Brittanica*'s *Propaedia*. The original sources, Plato's *Phaedrus*, *Sophist* and *Statesman* are interesting, as is just about any work by Aristotle.

Formative exercise

Look into the literature in some area with which you are familiar, and find some classification systems. You should not have to look very far.

- How large are they?
- Are they organized into hierarchies? How wide? How deep? Is the hierarchy balanced?
- Do they appear to be organzed by theoretical analysis? Literary warrant? Both? Some other?
- Do the class names at a level have names of about the same generality? On what basis do the subdivisions appear to be made?

Tutorial exercises

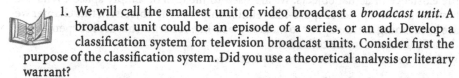

1. We will call the smallest unit of video broadcast a *broadcast unit*. A broadcast unit could be an episode of a series, or an ad. Develop a classification system for television broadcast units. Consider first the purpose of the classification system. Did you use a theoretical analysis or literary warrant?
2. Suppose this system is to be used to classify broadcast units for five channels over five years. What is the specificity of the various levels of classifier? Suppose it is to be used for one channel for one week. What difference does this make to the design of the system?
3. Consider a hierarchical portion of the classification system developed in Question 1. Is the hierarchy wide (hence shallow) or narrow (hence deep)? How is the hierarchy constructed?
4. Develop a classification system for the spikes of Appendix L.

Open discussion question

What are the advantages and disadvantages of a theoretical classification scheme over one based on literary warrant? Think of concrete examples, such as the spikes, the subjects in a university course catalogue, videos in a video shop, the Saturday or Sunday newspaper.

9 Semantic dimensions

We begin to see how classification systems can give us useful analogues of space, giving us tools to help in understanding and presenting information.

Place

We have already encountered the metaphor of space. In the information spaces we have seen so far, the primary aspect of space we have considered is that of nearness. We have seen that there are sensible definitions of how near one document is to another, based either on the presence of common terms (Chapter 3) or on explicit hypertext links (Chapter 5).

Probably a more fundamental aspect of the idea of space, however, is that a space is a system of places where things can be. This system of places is independent of the existence of actual objects. We are used to the idea of "empty space". What we have seen so far is a set of space-like relationships among objects. In this chapter we are broadening the metaphor to include place-like concepts that are independent of the actual objects in the collection.

Some examples of collections of places are:

- the squares of a chessboard;
- the rooms in a building;
- XY coordinates on a map;
- latitude and longitude on the globe.

In each of these cases, it is possible to go to a place whether or not there is something in it.

In the previous chapter we have examined the idea of classification systems. A classification system is a collection of names which is used to organize a collection of objects. Each object is assigned to exactly one (most specific) class. So, we can think of the classes as a collection of places. Furthermore, it is possible to have classes with no objects in them. For example, suppose we have developed a system of classification before assembling the collection, so there are no objects at all. All of the classes are therefore empty.

A system of class names behaves much as a collection of places. In this chapter we will develop this metaphor much more deeply, obtaining principles that will help us to design systems of classification that are useful for interpreting a collection of objects as a space.

What we can do with a place is look into it to see if there are any objects in it, and if so, which objects there are and how many. For example, we can classify the students at a university by their degree programme: arts, science, commerce, engineering, law, and so on. We can look at a class to get a list of the students in that programme, or we can aggregate the class to obtain, for example, the number of students enrolled in that programme.

Techniques of visualization can help us at individuals or aggregations, or often both. A classic graphic, shown in Figure 9.1, gives a history of the American involvement in the First World War. The system of places is the months from June 1917, when the first American units arrived in France, to October 1918, when the last units arrived. The graphic shows the units classified by month of arrival (the numbers are the identifiers of the units). It also shows what units were present in

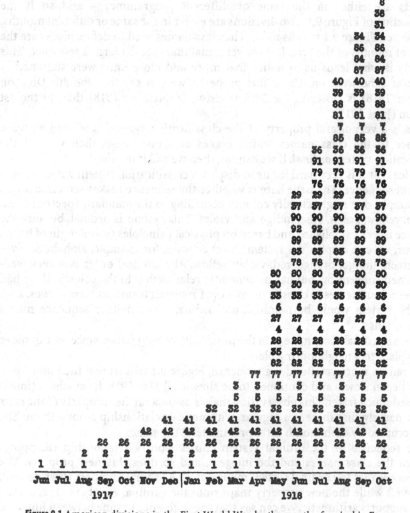

Figure 9.1 American divisions in the First World War, by the month of arrival in France.
Source: Ayres (1919: 102).

France during each month. Further, since no unit was withdrawn until after the end of the war in November 1918, the graphic shows the number of units present in any month.

Relationship among places

If we pursue the space metaphor, our existing concept of nearness among objects based on relationships among the content of the objects is somewhat strange. "Real" space has a concept of distance which is independent of the presence of any objects. What we have so far taken as distance is closer to the "real world" concept of similarity. We need nearness to be a property of places rather than objects.

Our simple place so far has a rudimentary distance – either two objects are in the same place, in which case their distance is zero, or not, in which case their distance is infinite. A list of students classified by degree programme works this way – students are either in the same or different programmes – and so is the classification of Figure 9.1. Two divisions are either in the same or different months.

However, Figure 9.1 tells us more. The class names used to define places are the names of months of the year. It is conventional that months form a sequence. This property is what leads us to notice that more and more units were stationed in France as time went on. Using this property, we can say that the 8th Division (October 1918) is closer to the 34th Division (September 1918) than to the 1st Division (June 1917).

This is a very useful property of the classification system. If we can assign a sequence to the class names which makes semantic sense, then we call the classification system *ordinal*. If we cannot, then we call it *nominal*.

Notice that it is very hard for us to display a classification system without some apparent sequence. The issue here is whether the sequence makes semantic sense. For example, one might classify colours according to the standard spectrum: red, orange, yellow, green, blue, indigo and violet. This system is ordinal, because the sequence is well established and based on physical principles (wavelength of light). However, we can display the system in other ways, for example, alphabetically – blue, green, indigo, orange, red, violet, yellow. Alphabetical order is a very well-established convention, but has no semantic relationship to the colours. If we had used the colour names to classify members of sporting teams (as team names, say), then the system would be nominal, not ordinal, since neither sequence makes semantic sense.

With an ordinal system, we gain the possibility of *navigation*, since we can move from a place to a neighbouring place.

We can do more yet. Besides a sequence, Figure 9.1 tells us that the build-up of troops began slowly and continued to be slow until May 1918, from which time it increased much faster. This observation makes use of a further property of the class names, namely that there is a regular arithmetical relationship among them. The time occupied by each month is (approximately) the same.

The sequence of the ordinal system comes from a relationship (mapping) between the class names and the integers, and so does this new property. The difference is that the ordinal system maps onto the ordinal numbers (first, second, third, etc.) while the new property maps onto the cardinal numbers (1, 2, 3, etc), which supports arithmetic. We can say that the difference in time between June and

September 1917, is the same as the difference in time between June and September 1918. We cannott say that the difference (in what?) between red and yellow is the same as that between blue and violet.

If we can assign the class names to the cardinal numbers, or the real numbers, or to any system supporting arithmetic, then we call the classification system *linear*.

With a linear system, we gain the possibility of *density*, because the places are organized in a system that permits arithmetic relationships to be computed.

Notice that the three properties linear, ordinal, nominal are nested. A linear system can be treated as ordinal (by ignoring the possibility of arithmetic), and an ordinal system can be treated as nominal (by ignoring the possibility of sequence).

Multiple semantic dimensions

When we think of space, we generally think of dimensions, and generally more than one. We live in a three-dimensional space. Edward Abbott's *Flatland* (1952), a two-dimensional world, is limited, but many human activities take place on a two-dimensional space (games like chess, field sports like soccer or hockey, most city planning, open plan offices). One dimension is very restrictive (Lineland, in *Flatland* , is a very strange place – people can't move past each other, for example). In mathematical systems, and indeed in databases, one often has spaces of many dimensions.

It is common to classify a collection of objects in several different ways. We can classify students by year and grade point average as well as by course. A sales analysis in a retail chain will often classify sales by location, by time and by product.

Following the theme of this chapter, that of use of classification systems to define spaces, we will call a particular system a *semantic dimension*. So if a collection of objects is classified in several different ways, then we will say that there are multiple semantic dimensions. Several of the examples in the previous chapter involve multiple semantic dimensions; in particular the noodle restaurant where there are three: *noodle, sauce* and *main ingredient*.

A place in a multiple semantic dimension classification system is an analogue of a box or a pixel.

Data analysis is very frequently carried out in information spaces of two semantic dimensions, often both linear. Consider Figure 9.2, where the nominal system is names of countries, which are used to classify their populations. The two semantic dimensions are per capita consumption of cigarettes in 1930 and crude male death rate for lung cancer in 1950. Both dimensions are linear, so a regression analysis can be done, as shown in the figure. The analysis shows a strong correlation between the locations in the two dimensions of the information objects.

But of course the dimensions need not be linear. A tabulation of university students by course and country of origin uses two nominal dimensions, while a tabulation by course and year uses one nominal and one ordinal dimension.

Many systems have more than two dimensions. Figure 9.3 shows a typical sales analysis schema, where the dimensions are store location, product sold and day of sale. The cells will typically contain the value of sales in some standard currency. Product and location are both nominal dimensions, while day of sale is linear.

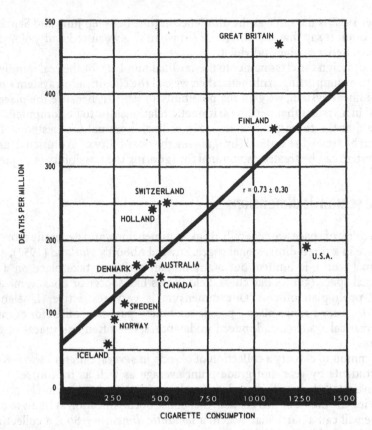

Figure 9.2 Crude male death rate for lung cancer in 1950 and per capita consumption of cigarettes in 1930 in various countries.
Source: Surgeon-General (1964: 176).

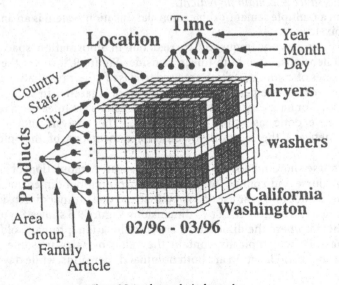

Figure 9.3 A sales analysis data cube.

Absolute and relative dimensions

Up to this point, all the classification systems considered have been exhaustive. This is not necessary, however. The three semantic dimensions in the noodle restaurant menu exclude the soups, so exhausts only the dry noodle part of the menu. The display in Figure 8.4 shows the most general system of classes: *wet noodle* and *dry noodle*. So the three-dimensional system applies only to the most general class, *dry noodle*. The other most general class, *wet noodle*, has only three subclasses and uses a quite different system.

This sort of situation occurs very commonly. Where the relevant population for a system of classification is not the entire collection but only the members of a single class, the system is called a *relative* classification system. So the *noodle, sauce* and *ingredient* dimensions are relative to *dry noodle*, while the {Laksan, Wonton, MeatBall} semantic dimension is relative to *wet noodle*. In contrast, a system which is exhaustive of the entire collection is called an *absolute* semantic dimension or classification system.

We have encountered many relative systems already. A hierarchical system like the SIC system, the *Encyclopedia Brittanica Propaedia* and the Yahoo! systems discussed in the previous chapter can be seen as collections of relative systems. The most general set of names in each case is exhaustive of the whole population being classified. These are the SIC Divisions, the Yahoo! top level, and the ten parts of the *Propaedia*. So those systems of names are all absolute.

The next level down in the SIC system is the major group. Notice that the names in the major groups differ depending on which division they belong to. So we can think of the names in Major Groups 01–09 as a semantic dimension relative to Division A, the names in Major Groups 10–14 as relative to Division B, and so on. Similarly, the second level systems of Yahoo! differ among the top-level classes, so that the semantic dimension relative to *Education* is different from that relative to *Computers and Internet*; and the divisions are different below each part of the *Propaedia*.

Hierarchical systems tend to involve relative semantic dimensions, but not necessarily. We have seen in Figure 8.4 that we can take several absolute semantic dimensions and turn them into a hierarchy by applying them in sequence. So in the *dry noodle* system, each different type of noodle is subdivided by the same set of sauce names, and each sauce name below each noodle name is subdivided by the same set of ingredient names. Each set of names is exhaustive of the whole population of *dry noodle*. It should be apparent that we can apply these absolute dimensions in any order to create six different hierarchies.

Systems of relative semantic dimensions do not necessarily create hierarchies. Consider a hypertext organized according to the concept map of Figure 5.4. We can use the names *historical antecedents, current practice*, etc. to classify links, and there may be many links belonging to each class. If we consider the set of links anchored in *current practice*, they are classified into the four classes indicated on the diagram, one of which is *possible future practice* and another is *problem to be solved*. Links anchored in *problem to be solved* are classified into three, including *current practice* and *changing problem environment*. Links anchored in *possible future practice* are classified into three, including *current practice* and *changing problem environment*. So each type of page has links which are organized into a system of classes. The system of classes differs for each type of link, so the systems are all relative. But the subclass systems are not hierarchical – in fact they form many cycles.

For another example of non-hierarchical relative systems, imagine you are in a major city with a subway network. You are at a station. You can classify the places you could visit in one train trip by which station they are nearest. This system is relative to the starting station. It is a classification system – each place is assigned to one station, but not hierarchical, since some of the places you can reach from your starting stations are other stations. The second-level classifications can include some of the same places as the first. If the rail network is complex, some stations are at junctures of lines, so the places accessible from some stations can be very different from the places accessible from others, yet a multi-train trip can reach any place from any other. This system is also cyclic.

But a non-hierarchical relative system does not need to be cyclic. Take a collection of topics taught in university courses. All university courses depend on assumed knowledge taught in other courses, possibly in secondary school. A particular course has a collection of assumed knowledge topics, which can be classified by the course in which they are taught. This system is not hierarchical, since the same topic may appear, possibly indirectly, as prerequisite knowledge for many courses. There is also no top, since no student would be expected to know all the topics taught in the university. We would expect no cycles, however – it is a poorly designed curriculum if two topics are prerequisites for each other.

All of the relative systems we have looked at so far are nominal: the noodle restaurant menu, the SIC, Yahoo!, the *Propaedia*, hypertext links, places by rail station, and prerequisite topics are all systems of names with no particular relationship among them. Relative ordinal or relative linear systems also exist.

Consider the subdivision of geological time into major geological intervals for the purposes of palaeontology. All evidences of life periods of time are classified into five eras: Cenozoic, Mesozoic, Palaeozoic, Proterozoic and Archean. This system of names is absolute, since it exhausts the population of evidences of life. It can be seen as ordinal, in that it is ordered by distance in time from the present. It can also be seen as linear, since with each name there is associated its ending time in millions of years before the present, also the starting time in the same units.

Each age is subdivided, however, into a similarly structured, but different system of periods. Similarly, some of the periods are subdivided. The whole system looks like Figure 9.4. The nominal systems are shown by indentation, the ordinal by the sequence of names down the page, and the linear by the end time in millions of years before the present. The system is relative, either ordinal or linear depending on your specific concerns, and hierarchical.

A second ordinal relative system might occur in a history of various branches of science. We could organize this hierarchically, with the top absolute level the names of the branches themselves. Below each branch would be a subdivision into historical periods that differ greatly from branch to branch. Mathematics would have subclasses for the work of ancient Greeks and Egyptians, while Computer Science doesn't start until the 20th century. The top level is nominal, and associated with each is a relative ordinal system.

The landscape provides a good example of a relative linear system which is not hierarchical. At each point one can classify what can be seen in three dimensions, typically *range* (how far away), *azimuth* (angle from the horizontal) and angle (*angle* from due north). Each of these dimensions can be represented at any appropriate granularity. Even though the relative dimensions at each point have the same names as at every other point, the system is still relative, since each system classifies different objects. The set of objects visible differs from point to point so

Cenozoic	0
Quaternary	0
Recent	0
Pleistocene	0.5
Tertiary	2.5
Pliocene	2.5
Miocene	7
Ogliocene	26
Eocene	38
Paleocene	54
Mesozoic	65
Cretaceous	65
Jurassic	136
Triassic	190
Palaeozoic	225
Permian	225
Carboniferous	280
Pennsylvanian	280
Mississippian	325
Devonian	345
Ordovician	430
Cambrian	500
Protoerozoic	570
Edicarian	570
Archean	2500

Figure 9.4 Major geological intervals, ended millions of years before present.

range, azimuth and angle are not exhaustive. This system gets its relativity not from differences among the sets of class names (they are all the same), but from the lack of a centre from which all objects are visible.

We now have a classification system with which to classify classification systems. It has two absolute dimensions {absolute, relative} and {nominal, ordinal, linear}. The first is nominal and the second ordinal.

Use of semantic dimensions for understanding a whole information space

Our three types of semantic dimensions give us three spatial properties: place (nominal), neighbourhood (ordinal) and density (linear). We now look at how these three spatial properties can be used.

The most basic property is place. We have seen that we can put things in a place, and therefore look at a place and see what has been put into it. We can look at either the individual objects or some aggregate, say the number or the average of some attribute. We get to look at a place by selecting its name, so the database operation of selection on class names identifies a collection of places to be examined. This activity is so common as to hardly need examples. We either make lists of object names grouped by class name or produce aggregations either as tables or graphs. We can also give properties to places. In particular, following Chapter 5, we can designate places as *explored* or *unexplored*, which is a two-valued absolute semantic dimension.

Ordinal dimensions allow us to see more. Figure 9.5 is derived from Figure 9.1. The arrival time of the units is classified into four classes: *early, start main, end main* and *late*. The *early* units arrived in small numbers over the first months. The main force arrived over three months from May 1918. The very large contingent which arrived in May is classed as *start main*, while the units arriving in the remaining two months are classed as *end main*. Units arriving after June 1918 are classed as *late*.

From this figure, we can see that the 80th Division arrived at the same phase as the 30th, but in a later phase than the 1st and earlier than the 36th. The layout also provides a visual aggregation, in effect a bar graph, which shows that the number of units being sent progressively increased in each phase.

Figure 9.5 does not show density, however, since the semantic dimension is only ordinal. A second re-drawing of Figure 9.1 showing the units arriving by month is shown in Figure 9.6. This figure shows the long slow build-up, then the rapid arrival of the main force, followed by large numbers of late-arriving troops. The additional information provided by the linear dimension lets us see more. As with the previous figure, the display shows both the individual objects and a visual aggregation.

Using the concepts developed in this chapter, we can see that Figure 9.1 actually shows two semantic dimensions. We have already looked at the linear dimension along the bottom. If you compare Figure 9.1 with 9.6, you can see that the former shows extra information, namely how long the unit was stationed in France, as an

Early	Start main	End main	Late
			8
			38
		36	31
		91	34
1	80	79	86
2	30	76	84
26	33	29	87
42	6	37	40
41	27	90	39
32	4	92	88
3	28	89	81
5	35	83	7
77	82	78	85
Early	Start main	End main	Late

Figure 9.5 US army units arriving in France during the First World War, by an ordinal classification.

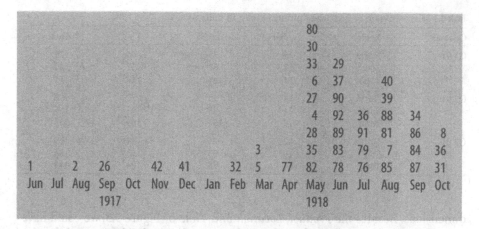

Figure 9.6 Variation of Figure 9.1 showing units by month of arrival.

unlabelled ordinal dimension in the vertical axis. There is one place in that dimension for each month in which units arrived, and the places are drawn in different sizes, so as to just fit the unit names, using the device of repeating the names of the units already there below the newly arrived units. This dimension is represented as ordinal, because there is no place for the months in which no units arrived, so no places for units stationed in France for 16, 13 or 10 months.

Month of arrival, the linear dimension employed, is quite coarse. The places it gives us are big enough to hold many units. We could have used a much finer time dimension, say the day, hour and minute of the debarkation of the first men of a unit, in which each unit would have only one place. We could then represent the increasing American involvement as a density of dots in a time line, or the cumulative force as a line graph. Using this representation, we would be unable to identify the individual units, except perhaps by zooming in using a dynamic visualization.

Our main problem with spatial representation of data is that the abstract concept of space used in constructing the system of virtual places is much richer than the physical space we can perceive. Our visual world is limited to three spatial dimensions, plus time. If we want to see the whole of a space, we are limited to two dimensions, since as we have already seen, the third spatial dimension yields a system of relative dimensions rather than absolute. We can never see all of a three-dimensional space. Time, by definition, is passing. If we use animation as part of our visualization we can see only one place (time slot) at once.

If we want to visualize a rich structure like the data cube of Figure 9.3, we have to give up visualizing some of the spatial relationships present in the data. We would generally present a tabulation something like that shown in Table 9.1. One dimension is represented horizontally, but two vertically.

The adjacency of the ordinal *Region* dimension is preserved, but the *Quarter* dimension is represented more abstractly, and is more difficult to see. The *Quarter* dimension is represented aggregated over *Region* as totals vertically, so their adjacency is a little easier to see. This sort of strategy can be extended to many dimensions, but the immediacy of visualization is lost.

This limitation in human perceptual ability means that complex systems must be visualized in multiple aspects. A data cube may easily have six or more

Table 9.1 Sales by region by quarter by product group, in million AU$.

Quarter	Region	Video	Audio	TV	Total
Q1 01	North	31	79	66	176
	South	22	62	51	135
	Total	53	141	117	311
Q2 01	North	53	53	72	178
	South	24	98	33	155
	Total	77	151	105	333
Q3 01	North	42	23	44	109
	South	19	53	62	134
	Total	61	76	106	243

dimensions. Besides the *Product*, *Location* and *Time* of Figure 9.3, we could have age, income and education of purchaser, for example, obtained from their loyalty card application. To visualize this data, we need to first select which dimensions we want to include as the main horizontal and vertical dimensions, then which other dimensions to include as nested subdimensions in each direction. The operation of selecting among multiple dimensions in order to choose an aspect from which to visualize a population of data is called *projection*.

We may not want to see the entirety of some dimensions. In Table 9.1 we have seen only three-quarters of one year, and only three product groups. So we can make selections of the possible places in any of the dimensions. In the data warehousing world, this selection of dimensions and subspaces is called On-Line Application Processing (OLAP).

Navigation in a hierarchy

In the previous chapter, we have seen that due to another limitation of the human mind, that most systems of classification are hierarchically organized. Many of these hierarchical organizations employ relative dimensions, while others employ nested absolute dimensions. In the data cube of Figure 9.3, two of the dimensions, *Product* and *Location*, are hierarchies with relative dimensions, while the third, *Time*, has three absolute dimensions. Every date is a year, every date is a month, and every date is a day of the month.

If we have a hierarchy, then a place can be subdivided, either by applying another absolute dimension (nesting) or by using a new dimension relative to the place. The process can be repeated until all absolute dimensions have been used or the leaf relative classes have no further relative subclasses. This process is familiar to anyone who has used a menu system. It is also possible to aggregate a place with other places formed from the other subclasses in the nested or relative system (going back up the menu).

In OLAP systems the process of subdividing places is called *drilling down*. The process of aggregating places is called *rolling up*. Someone examining Table 9.1 might be interested in the anomalously low sales for TV in the South region in Q2-01. They could drill down the region dimension to show the district totals. Finding perhaps a district with low sales, they might further drill down the product dimension to the types of TV. Finding that a new type of TV has low sales for that district in that quarter, they might drill down the quarter by applying the nested dimension week, giving weeks within quarter. This might show that sales dropped off markedly at week 7. Seeing this, the analyst might check for delays at the main port serving that district and find that sales were down because stock was unavailable. This successive drilling down, together with other evidence, now has explained the original anomalous sales figure.

Similarly, an analyst looking at sales of particular products in a particular city in the weeks of a quarter might see a fall-off in sales. They might roll up to the product type superclass and see that the sales pattern at that level was normal. They might then drill down again to the product level and roll up to the region level, to see that sales for this product were low for all the districts in the region. This might lead the analyst to check the press and discover that there had been an unauthorized announcement of a markedly superior new model intended to replace the model in question. OLAP is all about navigation through a complex data space by drilling down and rolling up to gain understanding of the situations described.

Visualization of relative dimensions

Visualization of systems of absolute semantic dimensions is relatively straightforward, since there are fairly straightforward methods of representing these dimensions in the real space of the page, computer screen, or other rendering device. Not so with relative dimensions, since each class may be subdivided on a different basis.

For hierarchical relative systems, a common way to visualize the information space uses a *dendrogram*, as in Figure 9.7. That figure shows the relationships among the languages of the Indo-European family. The top level, from Hittite to Baltic, is an absolute classification system, but each of these classes is subdivided on a different basis, which is why the system is relative but a hierarchy. The basis of the hierarchy is descent from a common ancestor, so that the whole system is intended to show the family relationships among the present-day or historical languages, often using hypothetical dead languages like Germanic or Celtic.

This visualization works because the whole space is small. We have noted many times the severe limitations on the ability of a human to comprehend a complex structure. In this context, the principle is reflected in the difficulty a human would have to take in many more than 100 classes at once. Figure 9.7 has a few more than 40. The Yahoo! menu shown after a fashion in Appendix B has fewer than 100 choices.

When the system is bigger than can be taken in, it becomes necessary to limit the number of branches that are represented at depth. It is very common in middle-sized systems for such a table of content to expand one choice at each level while leaving the other choices displayed on the screen. In very large systems, like Yahoo!, one generally displays the sequence of menu choices leading to the relative dimension displayed on any one screen.

The problem is quite different when the system of relative dimensions is not hierarchical. These systems are commonly represented as graphs, like Figure 5.4. The limitations on human understanding with which we have been working mean that only very small systems can be represented fully in that way. Small-to-medium Web sites often have site maps, but the practice does not scale to larger sites.

There are essentially two ways to deal with larger non-hierarchical relative systems. One is from the point of view of someone navigating the system. In this

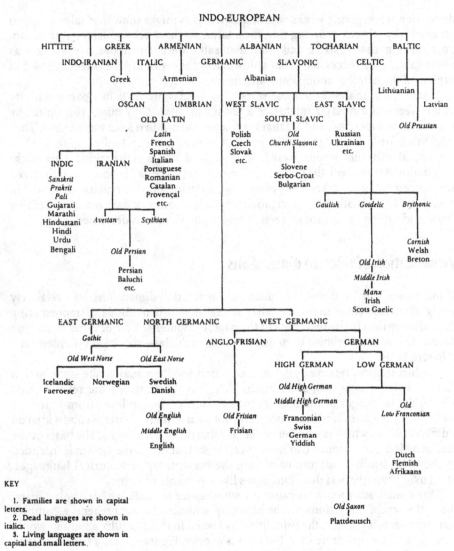

Figure 9.7 Family relationships among the Indo-European languages.
Source: Macquarie Dictionary (1987: 2008).

method, the user is presented with the dimension relative to the present point. This is the way hypertext systems are presented to the user, and the source of the whole navigation paradigm.

Alternatively, one can take an external view. The whole system is too big to visualize, and there is no place from which the whole system can be seen in an aggregated form (the system is not a hierarchy, remember). So any external view must be from the point of view of someone metaphorically above some point in the system. This leads to visualizations like Figure 5.5, a so-called *fish-eye view*, where the neighbourhood of the present position is shown in some detail, with progressively less detail shown for points progressively more distant.

Virtual reality systems are implementations of fine-grained relative ordinal or linear systems, making use of the metaphor of movement in real three-dimensional space. We return to this topic to a degree in Chapter 13, but the general issues of virtual reality are beyond the scope of this text.

Geometries based on many dimensions

We observed at the beginning of this chapter that the distance concepts of Chapter 3 were based on metaphors of similarity among objects, then contrasted this notion of space with a space notion based on relationships among places defined by the classification system. With the machinery we have now developed, however, we can see a relationship between the two views of space.

The relationship is based on the concept of *hamming distance*. In information theory, one is concerned with the possibility of corruption of digital messages by noise. If a message is represented as a string of ones and zeroes, then the effect of noise is to change some ones to zeroes and some zeroes to ones. Two strings are considered near to each other depending on the minimum number of errors needed to change one message into another. This measure is called hamming distance. The string '1011' has a hamming distance of 1 from both '0011' and from '1111', and a hamming distance of 2 from '0111' and '1000'. This measure makes engineering sense only when the messages consist of many bits, typically hundreds.

Document nearness query languages based on indexes, such as described in Chapter 3, can be regarded as geometries based on hamming distance. The index used for a collection of documents is derived from the words present (or possibly present) in the documents in the collection. The document vector model can be seen as a multidimensional absolute classification system, with one dimension for each term in the index. Each semantic dimension has two classes, say '1' and '0'. A document is classed '1' on the dimension associated with a given term if it contains that term, otherwise it is classed '0' on that dimension.

For example, documents 4' and 5' of Chapter 3 are represented in Figure 9.8 as a system with 20 nominal dimensions, each with two classes '0' and '1'. You can see that the hamming distance between Doc 4' and Doc 5' is 20, since they have no words in common.

This space depends on the terms in the index, irrespective of whether any documents in the collection actually contain a particular term. In fact, the collection could be empty. A point in this space is a string of ones and zeroes, one for each term in the index. The hamming distance between two points in this space is the same whether the points represent any documents or not.

Unfortunately, hamming distance is not a practical measure in a document space. A typical large collection is indexed by perhaps 500,000 terms, consisting of English words together with names of persons, places, organizations and acronyms. If the documents in the collection are abstracts, say, no document contains more than a few hundred terms. Therefore each document is represented by a point in the index space where nearly all of the dimensions have the class '0'. This means that the hamming distance between any two documents is relatively small, since even if two documents share no terms (as 4' and 5' below), they share the non-existence of hundreds of thousands of terms.

This is why the nearness measure of chapter 3 is based on shared terms only. The only dimensions considered between two points are those where at least one point

Dimension	Doc 4'	Doc 5'
beautiful	0	1
boat	0	1
criminal	1	0
existence	1	0
identity	1	0
known	1	0
merely	1	0
more	1	0
notorious	1	0
occasionally	1	0
often	1	0
organization	1	0
owl	0	1
pea-green	0	1
pussycat	0	1
rumour	1	0
sea	0	1
subject	1	0
vague	1	0
went	0	1

Figure 9.8 Space with multiple dimensions based on occurrence of terms.

has the class '1'. But the fact that in practice we use a variant of hamming distance does not invalidate the geometrical view of the index space. The distance between two points is defined whether or not any documents are represented by the points.

Hamming distance type measures on high-dimensional spaces can be extended to systems with more than two members of each semantic dimension class system. This general idea is taken up again in Chapter 12, where descriptors form a hamming-type space, and Chapter 13, where the concepts of semantic dimensions are used as the basis of a discussion on visualization.

Key concepts

One way to organize a classification system is as a number of **semantic dimensions**, each of which is a collection of descriptors which may or may not be hierarchical. The dimensions can be **absolute** (dimensions are independent) or **relative** (the content of one dimension depends on the value of another). The dimensions can be **nominal** (a collection of names), **ordinal** (the collection has an ordering), or **linear** (each name has an associated numerical value which is analogous to distance).

A hierarchical classification system (set of relative semantic dimensions) can be visualized as a **dendrogram** (tree-like diagram).

Semantic dimensions are used to visualize an information space. The visualization can be a **projection** of a many-dimensional space onto a few dimensions. If a dimension is hierarchical, the visualization can navigate in hierarchy – **drilling down** is seeing a more specific set of descriptors, **rolling up** is seeing a more general set.

An information space can be visualized either as a collection of objects or as **aggregates** (count, sum, average) of some property of the objects.

Further reading

 The concepts of semantic dimensions come from Wexelblat (1991). Gould (1993) has a table of contents which is a marvellous example of a visualization using multiple hierarchical linear semantic dimensions. More discussion on Figures 9.1 and 9.2 is in Tufte (1983).

Formative exercise

 Collect a number of visualizations of data spaces in areas with which you are familiar. Considering each:

- What semantic dimensions do they use?
- How do they fit into the taxonomy of this chapter?
- How are they represented in the visualizations?
- Are individual objects included in the visualizations, or aggregates?
- If individual objects are included, can you make an aggregation visually?
- Could the concepts of drilling down/rolling up apply to it?

Tutorial exercises

1. Consider the restaurants of your city.
   a. Who might want to know what about them?
   b. Develop a classification system which would support these requirements.
   c. Are any of the semantic dimensions relative?
2. How might the restaurants of your city be visualized to satisfy the various purposes from Question 1?
3. Develop a classification system to be used to support an On-Line Analytical Processing application where the individuals are university students enrolled in degree programmes.

Open discussion question

Consider a large term-based classification system derived mainly from literary warrant, such as the *Encyclopedia Brittanica Propaedia*. How could it be used to support a visualization of the total content of the *Encyclopedia*? What aspects of the classification system make visualization easier, and what aspects make visualization harder?

10 Classification in context

Classification is not the only way to understand information spaces, nor is it always the best way. In this chapter we look at how information interacts with classification systems, show that visualization approaches based other than on classification can be useful, and finally look at the historical origin of systematic classification in the scientific era.

Individuals and classes

We have seen the principles of classification, and several examples of classification systems of various sizes and types. Ultimately, a classification system is used to place individuals in classes. We begin this chapter with a consideration of this relationship between individuals and classes.

A classification system includes a procedure for assigning individuals to classes. This procedure usually takes the form of a set of rules. For example, consider dog breeds. Most countries have a system of clubs or associations which control the breeding and exhibition of pedigree dogs. A type of dog becomes a recognized breed if there are sufficient people interested in it, and its reproduction is stable. It is important that when collies have pups they look like collies. A breed is a classification – it divides dogs into examples of that breed and non-examples of that breed.

Now consider a given dog, Harry. How is it established whether Harry is a collie? First, Harry must have a pedigree – his parents must have been collies. Second, there is a system of features of the breed, and Harry must be within the tolerance range of each of those features. Even if both of Harry's parents are collies, if Harry is born without a tail, or with short hair, or with some other deficiency, Harry is unable to be registered as a collie.

Individuals are recognized as members of a biological species by a set of measurable characteristics – in many orders by the structure of the reproductive organs. An individual business is assigned a Standard Industrial Classification code (encountered in Chapter 8) according to a set of rules determining its primary business activity. A census assigns a person a primary place of residence according to a set of rules. A university decides whether a student is majoring in information science according to a set of rules.

Clear procedures for assigning individuals to classes are important to combat the principle of uncertainty. It is important that each person doing the assigning makes as close as possible the same assignments, and that the assignments remain stable over time.

On the other hand, people make great use of classifications in their day-to-day lives. They recognize the breed of a dog. They swat a mosquito. They go to the bank. They tell you where they live. They consider themselves to be majoring in information science. How do people make these classifications? Certainly not by applying the formal rules.

Those people old enough will say that Harry is a collie because he looks like Lassie (a character in a series of eponymous movies and television programmes in the 1950s and 1960s), or a white highland terrier because he looks like Wee Jock (a character in the 1990s television series *Hamish Macbeth*), or a bull terrier because he looks like my dog Buster, which has already been established to be a bull terrier.

When people say they have swatted a mosquito, they do not make a minute examination of its anatomy. They have killed an insect which hovers and whines like a mosquito. The bank is where they go to deposit their savings (even if to others it might be a post office). They say they live in Brisbane even if they own an apartment in New York and spend time there each opera season. They say they are majoring in information science because they are studying a programme similar to other students who are majoring in information science.

This is to say that informally most people assign individuals to a class to which a similar individual has been assigned. This process of assigning individuals to classes depending on similarity to known members can result in a complex network of individuals similar on a pairwise basis but with very little overall in common.

Consider bread. Common words like "bread" are classifications – they divide the world into instances of bread and non-instances of bread, and a very large number of individuals are in the class. All sorts of things are called "bread". Some bread is made from wheat flour, raised with yeast and oven baked. However, bread can be raised with chemicals ("soda bread"). It can be made with other grains, or even potatoes. It can be steamed, fried, or baked on a griddle (pitta bread, or chippattis). It would be extremely difficult to come up with a definition of bread that would include all the things we call bread and exclude all the things we don't call bread.

Think of all the things that are close to being bread, but are not. Think of pancakes (sourdough pancakes are yeast-raised, remember), crackers versus crispbread, steamed bread (e.g. Boston brown bread) versus dumplings, English muffins versus pitta bread, banana bread versus banana cake.

A similar study could be made for cake, for shirts, and many other common objects widely used in many cultures.

This point was made in a systematic way by the philosopher Ludwig Wittgenstein (1953), in his book *Philosophical Investigations*. He begins with the example of "game", which is a word like "bread". There are very many different kinds of things called games, but it is very difficult, if possible at all, to arrive at a definition which would include all the activities called "game" and exclude all those activities which are not. He goes on to say that abstract words are used as elements of everyday practice, which he calls "language games", and get their meaning from their use in these practices. It is generally practically impossible to define them.

So we have two different views of the world, of a collection of objects. On one hand, we have a formal classification system used to organize the information space; and on the other the informal network of more specific terms linked by pairwise similarity. This clash is the underlying reason for the principle of uncertainty. The imposed formal classification system is at odds with the informal way we have of seeing the world.

As a result of this clash, nearly all classification systems have to deal with difficult-to-classify individuals (this is another way of stating the principle of uncertainty). Even a system as small and well-established as gender: male, female has grey areas. As many readers would know, there are individuals who for biological or social reasons do not fit well into either class for certain purposes. Still, the system works well enough in the great majority of cases.

Larger systems have more grey areas; and the more specific the classifier, the more difficult-to-classify individuals one will encounter. Take the Standard Industrial Classification system of Appendix A. There are a great variety of business models among companies, and they frequently change. Banks may become investment managers or stockbrokers or insurance companies, either by growth or acquisition. They may downgrade activities, say by closing retail bank branches, or may divest altogether. A classic case (in another industry) is that the Singer Corporation sold its sewing machine business, which had made Singer a household name.

Companies tend to stay in the same general line of business, however. A diversifying bank may fit comfortably into the Financial Services classification, but become difficult to classify between the more specific classes Banking and Insurance. Similarly, people tend to move fairly frequently, but change metropolitan areas less frequently and country even more rarely. A census classification "principal lifetime residence" would find far fewer difficult-to-classify individuals at the level of specificity of country, more at the metropolitan area level, and even more difficult individuals at the specificity of suburb or postal code. The uncertainty at the last level may be so great that the classifier becomes useless.

If there is a clash, resulting in so much uncertainty, then why do we persist with the formal classification system? The short answer to that question is that this is the way our society organizes information, and it is difficult to imagine a scientific/industrial civilization which does not have organized classification. So we have to put up with the difficulties.

The longer answer is that there are in fact other ways to organize information which are becoming more and more practical as information technology develops. We will pursue some of these below.

Variability within classes

Classification systems are commonly used to organize information spaces. We have seen in the previous section why there can be difficulty in assigning individuals to classes (the principle of uncertainty). This section looks at the space from the point of view of the class. If we have a collection of objects all belonging to the same class, how similar do we expect them to be?

It is very common to represent a class by a "typical" or "normal" member. To many people, a collie is a dog that looks like Lassie. We might say to someone that they look like a movie star, or don't look like an accountant. This tendency, sometimes called stereotyping, has been part of the scientific culture since at least the time of Plato.

In practice, however, the members of most classes vary widely in several dimensions. We might think of manufactured goods as typical of classes – every Bic pen produced looks like every other one, for example. (Note that you never see the pens rejected by Bic's quality control, nor what happens when one of the machines malfunctions.) However, this uniformity in manufactured products dates only from about the middle of the 19th century. It takes an enormous amount of effort to maintain this uniformity in the face of the many sources of variability.

Consider a collection of blue objects, and the variation in their colours; or a collection of university courses; or a collection of university students, even within the same course. The variation is enormous. This variation is the ultimate source of the principle of uncertainty, but the variation in say the colours of blue objects is not simply that some blues are close to greens and fall in a grey area between blue and green. Most of the blue objects will be unequivocally blue, just different blues. The principle of uncertainty arises because of the imposition of class boundaries on varying populations – some individuals are close to the boundaries and are therefore difficult to classify reliably.

The first conclusion we can draw from this observation is that it is generally misleading to represent a class by a "typical" member. To get an idea as to what is included in a class one must see something of the range of individuals in the class. What we do from here depends on the purpose of the classification system.

If our aim is to assist a viewer to gain an overall feel for what is in a collection of documents, then a technique of presenting random samples of individuals can be valuable. This method is used, for example, in the American Memory collection of the United States Library of Congress on its Web site.

Often, however, the aim of the classification system is to enable either scientific study of the class or to formulate some sort of policy decision which affects members of the class. In these cases, the variability of the individuals in the class must be presented more systematically. Some examples of these more proactive applications are:

- Biologists may want to study a species. A single species is a very specific class in the Linnaean system, but may have a very large number of individuals.
- A business may want to understand the sales of a particular product in a particular time period, in order to get some idea of the medium-term future of the product. Again, the product may be a very specific class in the product system, but there may be a great many units sold in the period.
- A political party may be interested in the voting intentions of people in a particular electorate. Here, the classification system may be much shallower, but the number of voters in that electorate is large.
- A medical professional body might want to study the attitudes towards its profession as reflected in television medical dramas. Medical dramas are a specific class of television programme, but there are many series sometimes extending over several years, and each series has many episodes.

A common approach to the systematic study of variation in a population of a class is to introduce subclassification using a number of semantic dimensions which are absolute with respect to the class of interest. For the biologist, the subclassifications might include size, colour and climate measured for each specimen. For the business, they might include size of the store from which the product is purchased, quantity purchased per customer per purchase, average socio-economic status of the area in which the store is located, and predominant linguistic or cultural characteristics of the people living in the area where the store is located. The political party may subclassify by age, sex, length of residence and nominal party affiliation. The medical profession may subclassify the dramas by degree of technical realism, amount of humour, average ratings of the series, and specific occupation of the central characters (hospital doctors, general practitioners, police forensic practitioners, nurses, psychologists).

The population will be tabulated by each of the dimensions. Where the dimension is linear, its variation may be represented by mean and standard deviation. The population may be cross-tabulated on two or more dimensions at the same time, as in the data warehousing example of the previous chapter.

One of the earliest studies of the variation of human behaviour, and still a landmark in the methods necessary to get a good picture, is the series of studies of sexual behaviour published in the 1940s and 1950s by Alfred Kinsey and his colleagues. *Sexual Behaviour in the Human Male*, published in 1948, devotes more than 50 pages to the semantic dimensions used and the sampling and statistical methods necessary to give reliable results.

His study was limited to white American and Canadian males. The objects in the information space are sex histories obtained by interview containing hundreds of items. He used 12 semantic dimensions to study variation: sex, race-cultural group, marital status, age, age at adolescence, educational level, occupational class, occupational class of parent, rural-urban background, religion, degree of adherence to religion, and geographic origin. He worried about the number of samples needed to achieve breakdowns by up to six dimensions taken simultaneously, and the variation in characteristics among the populations of the resulting cells.

This sort of many-dimensional cross-tabulation is becoming common today in the field of data warehousing. Such a data structure was called a *data cube* in Chapter 9.

In situations where the population of the class of interest is smaller and fewer semantic dimensions are used to describe its variability, the cells are often supplemented by random samples of original data records.

An example may help all this make sense. Let us assume we are interested in an aspect of student performance in courses. University Q has about 4000 courses which are grouped by department (about 60), by semester (two semesters), by student-year (1–4), and by calendar year. (That is to say, the courses are classified by four semantic dimensions: the first nominal and the remaining three ordinal.) The courses themselves form a relative system within each cell (there is no relationship between the course codes from one cell to another). The particular aspect we are concerned with is the performance in a specific year, say 2000, in a specific course, CS272, which is a course in information science taught from this text. In the year in question, 85 students completed the course.

One way to represent the students in this course for 2000 is by the average mark received (64.5%). This average mark does not convey very much information, since the University likes to calibrate the assessment in its courses so that the average mark comes out at about 65%. The assessment in CS272 is made up of four components, however, so more information is conveyed if we represent the class by the average performance on each of the assessment components, as in Table 10.1. We might think of this as the average or typical student. We can see that the average student performed best on component 3 and worst on component 4.

In the discussion above, we made the point that there is a great variability in most populations. The average student performance does not show that variation. So we can get a better picture if we show variation. A crude way to do this is to show the

Table 10.1 Average student in CS272 – 2000.

Component	1	2	3	4
Average	76	71	80	53

maximum and minimum marks for each component. This is preferable to more sophisticated calculations of variance or standard deviation as the students in this course are not expected to have much understanding of statistics, so may find stronger measures of variation difficult to interpret. Table 10.2 shows the variation in average student performance. There is a very large range in all the components, with the whole range for component 4 moved down considerably from the others.

It happens that CS272 is taken by students from a variety of degree programmes, which differ greatly in the style of thinking developed. The breakdown of students by degree programme is in Table 10.3. Note that the range is from arts, through education, business, computing, science, to engineering. There were a few students who were primarily studying at universities overseas. This list is ordered from the prominence given to qualitative thinking (arts) to prominence given to quantitative thinking (engineering), and thus can be thought of as an ordinal semantic dimension.

One might expect that students from different degree programmes might perform differently in a course, so we can represent the class by average overall mark with maximum and minimum, as in Table 10.4. We see that the engineering

Table 10.2 Average student in CS272 – 2000, with variation.

Component	1	2	3	4
Average	76	71	80	53
Max %	100	100	100	95
Min %	45	30	40	10

Table 10.3 Breakdown of CS272 – 2000 students by degree programme.

Course	Percent
Arts	28
Education	4
Business	18
Computing	15
Science	16
Engineering	14
Overseas	5
Total	100

Table 10.4 Performance of CS272 – 2000 students by degree programme.

Course	Ave%	Max%	Min%
Arts	63	89	16
Education	66	76	50
Business	69	88	53
Computing	59	76	35
Science	65	83	35
Engineering	70	90	55
Overseas	70	75	64
Total	65	90	16

and business students did best, while the computing students did worst. This result says that the hypothesis does not hold up that the difference in performance is explained by a (very loose) indicator of qualitative versus quantitative thinking.

Alternatively, since the course is an elective for most students, the students differ greatly in their degree of commitment to it, in their social connection with other students, and in their overall self-organization skills. One of the assessment components was done in groups organized early in the semester who made regular submissions throughout the course. Since the students were required to form groups themselves, it took several weeks for all the groups to be finalized. There were 20 groups in the end. We might guess that the students in the first ten groups formed (early) were perhaps more committed, better connected or better organized; and those in the last ten groups (late) were less on these qualities. This gives us another semantic dimension with two classes. Table 10.5 gives a breakdown of the class on this new dimension.

We can look at performance in these new subclasses, in Table 10.6, which does give support to the hypothesis that the students forming groups earlier are better organized and do better.

There are enough students in most of the degree programme subclasses to further subdivide by early and late. Table 10.7 gives a breakdown of students in degree programmes other than education and overseas by early and late.

And, finally, Table 10.8 shows average marks for each of the cells in Table 10.7. We see that the earlier organized students did better than the late for all programmes except computing.

Table 10.5 Breakdown of CS272 – 2000 students by early and late.

Group formation	% of class
Early	48%
Late	52%

Table 10.6 Performance of CS272 – 2000 students by early and late.

Group formation	Ave %	Max %	Min %
Early	68	90	16
Late	63	81	35

Table 10.7 Breakdown of CS272 – 2000 students by degree programme and early/late (% of total).

	Arts	Business	Computing	Science	Engineering
Early	12	7	5	11	9
Late	16	11	11	6	5

Table 10.8 Breakdown of CS272 – 2000 students by degree programme and early/late.

	Arts	Business	Computing	Science	Engineering
Early	64	73	53	68	74
Late	62	66	62	58	63

These various subclassifications of the students in CS272 increase our understanding of the variation in performance in the class, and enable us to generate and partly test hypotheses explaining this variation.

Classification is not the only way to organize information spaces

We have seen that classification is a widely used method of organizing information spaces: biologists use the Linnaean system, a large hierarchical nominal system; chemists use many smaller classification systems (acid, base, hydrocarbon, etc., or the periodic table of elements); and economists use the Standard Industrial Classification in Appendix A. In many cases, however, classification is not the best way to organize the space.

The evolutionary biologist Stephen Jay Gould makes this point in a number of articles. Biologists were among the first users of organized classification schemes, and these schemes have been spectacularly successful. Biological species are grouped into genera, families, classes, orders, phyla, kingdoms, and so on. A biological species is not a single entity, however. It consists of a possibly very large number of individuals occupying a possibly large number of different habitats, so many species have a great deal of variation among their members. Biologists studying a particular highly variable species often want to make sense of this variability.

According to Gould, until fairly recently biologists generally made use of the nominal hierarchical classification systems so successful above the species level, by developing a system of subspecies or races to explain the variability within a single species. Gould cites a case where a highly variable species is divided into 600 subspecies (a species of land snail which occurs only on a group of islands in the Carribean Sea). He argues as a biologist that it is difficult to get an overall view of the variability of the species from such classification systems.

He argues that for many purposes a much more productive approach is what he calls multivariate analysis. Instead of using a single hierarchical nominal classification scheme, he advocates use of several independent semantic dimensions, linear if possible or else ordinal, often highly specific.

For example, Gould cites a study of the English sparrow in North America. The species was introduced in the 19th century, and has spread throughout the continent. At the same time, the species has developed many local variations. A 1964 study declined to introduce subspecies, instead employing a number of semantic dimensions, including geographical latitude and longitude of the point of collection of a specimen and many specific numerical (hence linear) measurements on each specimen.

The method of analysis of this data was visualization, as in Figure 10.1. Here, the authors have selected three semantic dimensions. Two are latitude and longitude, so that the visualization is a map, and the third is a composite measure of size which is represented as an ordinal dimension with ten values. Size is represented as darkness, so that the largest birds are found in central Canada and the smallest on the California coast. (The 739 data points have been smoothed using standard techniques similar to those used to make contour maps or weather maps.)

Knowledge of the climate of North America leads one to make the hypothesis that the size of the bird is dependent on how cold the weather gets where the bird lives. We can introduce another linear semantic dimension, say average minimum

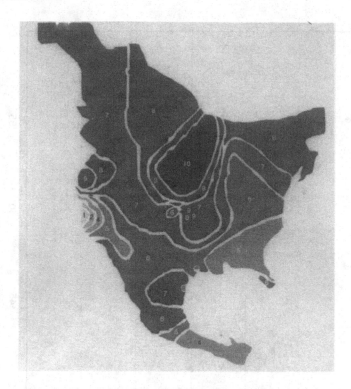

Figure 10.1 Size of sparrows by geographical location in North America
Redrawn from Gould (1977: 235).

January temperature at the latitude and longitude where the specimen was collected, to represent how cold it gets. The data can now be displayed in the two abstract dimensions, coldness and size, as what is called a scatterplot, to test the hypothesis. Such a display is given in Figure 10.2. This shows that, with the exception of a couple of outlier points, there is a close relationship. (The climate dimension is a composite called "isophane" that is large when the climate is very cold and small when very warm.)

There have been a large number of statistical and visualization techniques developed, which are only practical with the assistance of computers, that can be used to help understand variation without making hierarchical nominal classification schemes.

We can make use of multivariate analysis in understanding the variability of performance in the information science course which was analysed by classification in Tables 10.1–10.8. Figure 10.3 shows two semantic dimensions, total marks per student, represented as an ordinal dimension (class rank), and programme, represented as a nominal dimension (or perhaps ordinal, as mentioned above). Each student is represented by the combination of course and class position. Visualizations of this complexity require careful study. This visualization shows that there is huge variation in performance among students in all programmes. However, there are no top-performing students in the computing programme, and no bottom-performing students in the engineering programme.

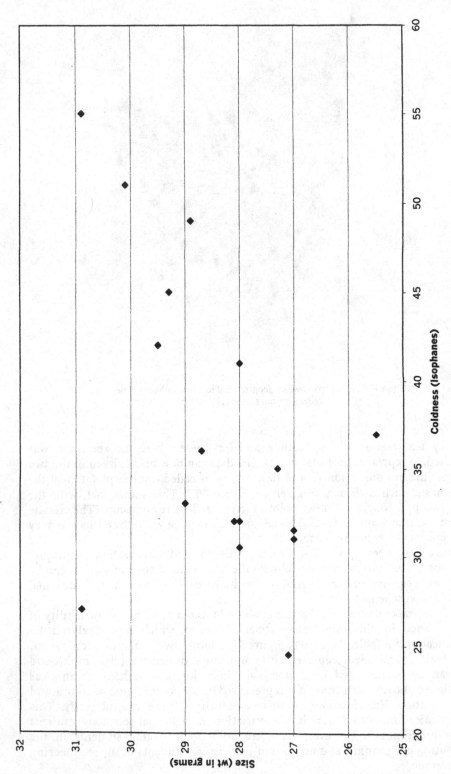

Figure 10.2 Size versus coldness of climate for sparrows in North America.
Source of data: Johnson and Selander (1964: 549).

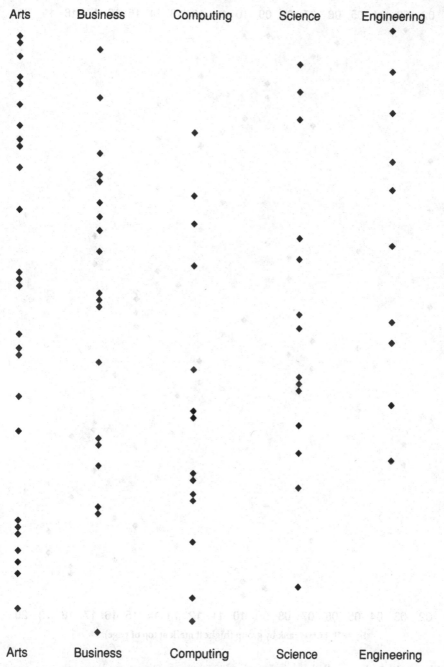

Figure 10.3 Class rank by course (highest at top of page).

Figure 10.4 shows a similar visualization using group number and class position. It shows that there is a wide variation in performance among even the members of groups. The first few groups formed have no very weak performances, and the last quarter of groups formed no strong performances.

Figure 10.4 Class rank by group (highest mark at top of page).

Figure 10.5 shows all three semantic dimensions. Group and programme are represented as vertical and horizontal coordinates, respectively. The third semantic dimension, performance, is shown by letters within the resulting cells. Because it is difficult to make fine distinctions in this type of representation, what is shown is not class rank, but grade (an integer between 2 and 7, where 2 is a failure, 4 is a pass, 5 is average, and 7 the highest mark. (The design considerations underlying this choice are discussed in Chapter 13.)

Group	Arts	Business	Computing	Science	Engineering
01		X			XXXx
03				Xoo	X
04		oo	o		
05					
06	X	X		X	
07	XXxO				x
08	x		x		x
09	XXO	Xx	o		
10	O	x	O	X	
11	X		Xxo	x	
12	XX	xx	x		
13	x		O	xo	
14	X	xo		Xx	
15	xxo	xx			
16	xO	o			
17	o	x	x	oo	
18	O		o	oO	
19	oO	x	o		
20	o	x	oo		

Figure 10.5 Grade versus programme versus group number.
Legend: $X = 7$, $X = 6$, $x = 5$, $o = 4$, $O = 3$, $O = 2$.

The figure shows that the arts and computing students tended not to form groups early, while the science and engineering students did. However, students from all programmes are represented in the late-forming groups. Students with the highest grade (X) tended to form groups early within their programme. There are many more below average students (o, 0, 0) from group 10 on.

Classification and multivariate analysis show the variability of the population in different ways. Classification in this case is more compact. Table 10.4 makes much the same point as Figure 10.3, namely that students from some programmes performed better than others, but the visualization shows much more data than the aggregations of the classes. The visualization shows that the definiteness implied by a uniform set of numbers in the classification is not a good representation of the data. All the programmes show great variation, and only computing and engineering show marked differences from the others.

Table 10.6 showed that early groups did better than later, but the corresponding Figure 10.4 shows that performance was better only for the earliest groups and generally worse only for the latest-formed groups. The division into early and late is shown to be arbitrary, and not a good classification.

Table 10.8 showed that for most programmes the early students performed better than the later, except for computing students. Figure 10.5 shows that the relatively strongest computing students tended to form groups in the middle range, from group 08 to group 11, but the fact that there were few computing students in the early groups and that one of them did very poorly (0) shows up in the table as the anomalous poor performance of early computing students.

Taken together with some domain knowledge, multivariate analysis cannot only describe the variation, but also demonstrate explanations for it. The observation in Figures 10.1 and 10.2 show a correlation between size of sparrows and the coldness of the climate. This correlation could be explained in at least three different ways:

- Coldness produces larger birds (it is commonly observed that warm-blooded animals tend to be larger in cold climates than in warm, and this observation is supported by physical and physiological explanations).
- Larger birds migrate to colder climates (or smaller birds to warmer). This relationship is observed in many human populations, where people living in cold areas move to warm areas when they retire.
- Coldness and size are both produced by some other, unobserved attribute. This alternative is unlikely in this situation, but consider the relationship between world records in track and field events and the sex of the competitor. The observation that performances in men's events are better than performances in comparable women's events is very likely caused by the fact that men are on the average bigger than women. In fact, a woman with a comparable size and build will generally perform about the same as a man. The observation is explained by the fact that the best male competitors are larger than the best female competitors.

Taking into account the knowledge that climate in an area does not change much from year to year and that sparrows tend not to migrate very far, we can interpret Figures 10.1 and 10.2 as indicating that coldness of the climate of an area is responsible for the increased size of sparrows living there. In Figure 10.1 the observed variation in size by location is caused by the third dimension, coldness, which is not included in the visualization.

What we do when we make explanations like that is to classify the semantic dimensions into exogenous and endogenous. An *exogenous* dimension represents an attribute which is taken to be the cause of the variation, while an *endogenous* dimension represents an attribute which is caused to vary. In the sparrows case, coldness is responsible for size.

In Figure 10.3, the observed difference in performance between students from different courses is likely to be produced by other attributes of the students – for example, stronger high school students have a tendency to enrol in engineering or business rather than arts or information technology. Alternatively, the relatively poor performance of computing students might be explained by the fact that the course is a somewhat peripheral elective and therefore not attractive to the better students. These hypotheses suggest that further investigation using other semantic dimensions might identify exogenous variables, which would explain the relationship between the endogenous variables shown in the visualizations.

What is in the information space?

So far in this chapter, we have argued that classification systems are tools established by humans to make sense of aspects of the world. As tools, they are subject to design, hence the design guidelines and methods of Chapters 7, 8 and 11. We have also argued that the choice of classification as a tool is itself subject to design – there are other tools which can be more suitable in particular situations.

We turn now to the issue of what should be in the information space in the first place. In many cases, of course, the document collection is given beforehand, but the question remains of what aspects of the documents should be included in the surrogates and the classification system.

We have touched on this question in the previous sections, without acknowledging it. In trying to understand the variability of results in our information science course, we established a subclassification system using students' degree programme and the number of the project group to which the student belonged. There are many other available characteristics which could have been chosen: students' age, sex, home residential postcode, and so on. The choice of course and group was not accidental.

The reasons mentioned in the previous section were that different degree programmes tend to develop different ways of thinking; that different programmes attract students of different average abilities; and that students forming groups earlier might have better time management skills than students forming groups later. Besides these, there is a political reason. University study is organized around degree programmes, and the course is an elective in several. Differential results from students in different programmes can be useful in negotiation with programme coordinators and in attracting students.

In general, we select the population of our information spaces and the way we describe them in ways that are strongly affected by our purposes and our theories. Stephen Jay Gould reports on the 17th century origins of modern geology in his essay "The Titular Bishop of Titopolis". There are lots of interesting things about rocks and minerals: crystal formations inside rocks, crystal formations that look like letters of the alphabet or pictures of significant people, shells and bones inside rocks, rocks with holes in them, and so on. Many people had made collections of interesting rocks, and had developed systems of classification to organize them.

The work reported by Gould is a book by the Danish scientist Nicolaus Steno who lived and worked near Florence. Steno had a theory about how rocks were formed, and developed a classification system based on this theory, which has been further developed and revised by geologists since then. Significantly, however, Steno excluded from consideration properties of rocks that did not correspond with his theory. In particular, he included crystal formations inside rocks, and fossil shells or bones found inside rocks, but excluded formations that by his theory only accidentally looked like letters or pictures, or odd-shaped rocks such as rocks with holes, the shapes having been produced after the rocks were formed.

According to Gould, modern geology owes much to Steno's theories, and possibly even more to Steno having excluded those phenomena which could not be accounted for by them.

For a further example, we turn to a map of an area of London affected by a cholera outbreak in 1854, made by Dr John Snow. The map shows the residential addresses of people who died in the epidemic (by a dot) and the location of water pumps (by an x). Notice that there is a well just at the centre of the map (near the 'd' of Broad Street) surrounded by dots indicating victims of the epidemic. The other wells are at the periphery of the concentration of deaths. Snow had a theory that cholera was caused by contaminated water, and used this map to convince the authorities to close the pump. The successful ending of the epidemic led to the founding of public health as a concern of government.

The cholera map, in Figure 10.6, has three semantic dimensions; two of which are geographic coordinates overlying the map. The third is represented by the marks on the map, which are of two types: the location of a death from cholera or the location of a well (a nominal system with two elements). We have seen that the map is a very strong argument that the cause of cholera is contaminated water, specifically in this case, the well in Broad Street.

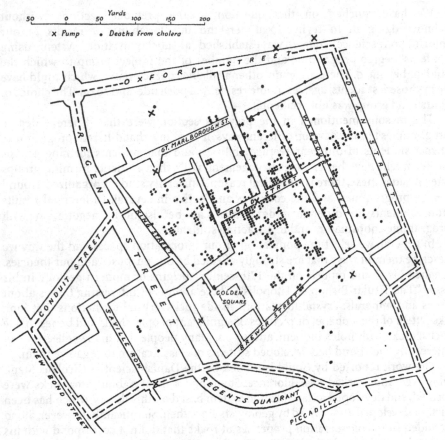

Figure 10.6 Cholera map of London, 1854.
Source: Gilbert (1958)

However, consider what is *not* on this map. The people who died from cholera differed in age, sex, occupation and nationality. There would have been a great variety of shops and barrows selling all kinds of food and other goods. There would also have been many places of work, with all kinds of trades. Why display only the bare death and residential location, and only the locations of wells? Dr Snow had a theory about the cause of cholera, and he used that theory (contaminated drinking water) to select the data to represent. His theory was successful, so that the map has gone down in history. Other people might have made other maps based on other theories, but, unsuccessful, they would not have been remembered.

The choice of what to include is a design choice.

Systematic classification is an aspect of modern culture

We have seen that classification is widely used in all sorts of activities. We complete this chapter with the fact that systematic classification was invented with modern European-derived culture, and that classification is pervasive in that culture, of which this text is a tiny fragment.

That systematic classification is recent is documented by the French philosopher Michel Foucault, mainly in his work *The Order of Things: An Archaeology of the Human Sciences* (1970). We noted in Chapter 8 that classification was invented by the Greek philosophers Plato and Aristotle in the 4th century BC. However, it is only one way of organizing knowledge. Foucault shows that in European society up until the end of the 16th century, the predominant method of organizing knowledge was the method of resemblance, where various things were connected based on similarities.

It is difficult to give a brief account of the system other than by a series of examples taken out of context which seem quaint or silly, whereas the system was very extensive and taken very seriously. It persists in contemporary society in a number of areas, notably the interpretation of dreams, where objects or situations dreamed about are taken to represent objects or situations which are problems for the patient, and which the dream object resembles in some way. It also appears in the metaphor approach to human-computer interface design, where collections of computer files are represented by images of manilla folders, deletion by an image of a trash can, and colour by an image of a paint tin. Anything can be related to anything else so long as some aspect of the first thing resembles in some way some aspect of the second.

Resemblance leads to a network of relationships among objects very similar to the relationship among abstract words described earlier in this chapter epitomized by Wittgenstein's analysis of the concept "game". In the same way that the informal use of abstract words leads to a network of local relationships among more concrete meanings, the system of resemblances leads to a world whose ordering principle is local – each thing is connected to other things by specific relationships.

Foucault shows that systematic classification based on global principles became common in scientific study in the 17th century. He also argues in a number of works (*History of Sexuality, Madness and Civilisation, Discipline and Punish*) that classification became the fundamental principle of social organization and control from about that time. It is essential to the organization of the modern bureaucratic nation state and to the large corporation. It is also essential to social control. Sexual behaviour is regulated first by the church and then by the medical and justice systems. Taxation is based on classes of goods and classes of services. Things that various kinds of businesses are permitted, required or forbidden to do are organized by systems of classification. Social and political rights and obligations are organized by classification. Is a convicted murderer allowed to vote? Can someone who has been certified as mentally incompetent make a will?

Imposition of a system of classification on previously unregulated behaviour will change that behaviour, often in unexpected ways. Introduction of a tax that taxes food for humans less than food for dogs will result in butchers selling "soup bones" rather than "dog bones". Differentiation between income and capital gains, with capital gains taxed less, results in tax minimization schemes through which activities previously considered earning income are re-labelled and re-organized as activities increasing capital value.

Alternatively, introduction of a classification system intended to change behaviours often has less effect than one might think. When an AIDS epidemic changes the priorities given to types of medical research, a research programme in cell physiology can become anti-virus research instead of looking for a cure for cancer, without changing its ongoing activity. When artificial intelligence falls out of favour, research into statistical estimation of classification rules can become data mining with hardly any change.

The point here is that classification as a method of organizing information is a design choice at a societal level. There have been other preferred methods, and there will likely be others in the future. Furthermore, the use of classification for regulation of behaviour requires significant effort in administering and enforcing the system – this is what bureaucracies largely do. Use of a controlled vocabulary in indexing a continuing stream of documents is an example of regulation of social behaviour. This is why controlled vocabularies are expensive to operate.

Key concepts

This chapter has covered some perhaps strange territory in its examination of the context of and limits to classification. It is intended to leave the reader with the following message:

- The world is weirder than we can imagine.
- We need tools to cope with its variety.
- Classification is one such tool,
  - but not the only tool.
  - nor always the most suitable.
- Classification systems are not discovered, but designed.

This final point, that classification systems are designed, not discovered, leads us into the next chapter where we look at design principles for large classification systems.

Further reading

Abstract terms as families of more concrete terms is the basis of Wittgenstein (1953). The Kinsey report is Kinsey et al. (1948). A short discussion of the report, bringing out the issues raised in this chapter is in Gould (1985: 10). Alternatives to classification are discussed by Stephen Jay Gould in Gould (1977; 1985: 11). The Steno story relating to the choice of what to represent is described by Gould (1983). The cholera map is described by Tufte (1983) and discussed in some detail in Tufte (1997). The most relevant work of Foucault is in Foucault (1977; 1986).

Formative exercises

1. Consider a class with a large population with which you are familiar. Is there a typical member? How much variability is there among the individuals in this class? How is this variability typically represented? If there is little variability, why not?
2. Consider a published report on a complex data set in an area with which you are familiar. How does the author present the data? By a classification system? By

visualization? Are there exogenous and endogenous variables? Are multivariate statistical techniques used? What could the author have included in the data but did not? Why was the data selected as it was?

3. Consider a system of regulation recently imposed on a domain of behaviour with which you are familiar. What classification system is used to support the regulation? What effect did the classification have on instances of behaviour?

4. Consider a system of regulation recently removed from a domain of behaviour with which you are familiar. What changes were there in the behaviour formerly regulated? Did new forms of behaviour appear? Did previous forms disappear?

Tutorial exercises

1. Think of as many ways as you can to classify university students. What purpose does each serve? Are any of the methods relevant to all purposes?

2. Consider how to understand the population of journeys from home to the university campus. Try both classification and multivariate analysis approaches.

Open discussion question

Take a general word with which you are familiar (like "bread" or "game"), and find as many diverse examples of the word as you can. Identify concepts near these which are not examples of your word. Try to define your word in such a way that all your examples satisfy the definition and none of the non-examples do. How easy is this?

11 Large classification systems

We have looked at the basic principles of controlled vocabularies, especially classification systems, in Chapter 8. In Chapter 9 we saw that the design of a classification system is influenced by its possible use as semantic dimensions in visualization of an information space. In Chapter 10 we saw that classification systems are engineering objects designed to solve problems, rather than scientific objects which can be discovered; and further that many problems are better solved by tools other than classification. In this chapter we look at some large classification systems and see how they are designed.

Introduction

The small to medium classification systems we have used as examples so far take some thought to develop but can be designed by a small group of people in a relatively short time. There are in the world several very large classification systems which have engaged the effort of many people over many years. The size of these systems introduces additional design problems.

Largest of all is the Linnean system of nomenclature used for biological species. This system has been under development for more than 200 years, and has millions of most specific classes. Thousands of scientists all over the world are engaged in its maintenance and development. Somewhat smaller, but still very large, are the systems developed by librarians for the classification of books and other resource material. Systems like the US Library of Congress classification system (LOC) and the Dewey Decimal Classification system (DDC) have been developed over 100 years by substantial teams, and are used by thousands of libraries. More recently, some industries have developed large classification systems – in particular the health care industry's SNOMED or Read Code systems used for classifying health care incidents has hundreds of thousands of classifications.

These classification systems are so large as to require classification systems to classify their class names. They are supported by large scale indexing systems.

What an indexing system involves

There are three main aspects to the indexing systems supporting the large classification systems.

Most obvious are the classifier names themselves, the semantic content of the system. In the library world, this is called a *schedule*.

Secondly, large indexing systems often have a *notation*. The library systems like LOC or DDC are used for organizing libraries, so the classifications are expressed in a notation for physical placement of objects – the book's call number. Other systems like SNOMED or Read Codes are used to classify medical records, so have a compact code for inclusion in the data records.

Thirdly, large indexing systems have their own *indexes*. Humans do the coding, so they need indexes to find the right codes to apply to particular objects. Humans do the searching, so need indexes to find the right codes to express their information needs. Remember that in classification, each object is normally assigned a single classification, so these indexes are essential to minimize the impact of the principle of uncertainty.

Requirements for schedules

The schedule for a classification indexing system is the set of class names. There are a number of design criteria which must be taken into account in schedule design. Some criteria are what are called *external* – they relate to the task the indexing system is required to perform. Other criteria are what are called *internal* – they relate to the quality, usability and maintainability of the indexing system. We will first list the criteria, then show how they are applied in the design of the Library of Congress System and the Dewey Decimal System.

External criteria include:

- Exhaustivity: as discussed in Chapter 8, the set of class names must be exhaustive. Each object in the collection must be able to be assigned to a class.
- Coding efficiency: space in schedule proportional to size of literature. In the language of Chapter 8, each most specific classifier should have approximately the same specificity.
- Semantic coherence: order in schedule reflects nearness in semantic space. The class names should be ordered in the schedule so that names that are near each other are used to classify objects that are semantically similar.
- Maintainability: the class names should allow for expansion and contraction of the collection of documents (the LOC and DDC systems have evolved over more than 100 years). They should also allow for change of relationships and interdisciplinary studies.

Internal criteria include:

- Each simple subject must have a clear place in the schedule.
- Each complex subject must have a clear place in the schedule.

Library of Congress system

Exhaustivity

An extract from the LOC system is shown in Appendix D. Observe first that the criterion of exhaustivity is satisfied by a hierarchical decomposition. Every book published must be able to be assigned to one of the most general classes, A *General Works* through Z *Library Science*. Notice that the first class, *General Works,* is designed to take books that do not fit into any other class. Each classification is then subdivided, with each subdivision being exhaustive within its more general class. Notice that classification K *Law* has a subdivision 181-184.7 *Miscellany*, which can be used for books in the class which do not fit into any other of its subclasses.

Coding efficiency

The criterion coding efficiency is shown by the varying number of levels of subdivision. For example, observe the classification Q *Science*. One of its subdivisions is QA *Mathematics*. It has nine subdivisions, from 1-43 *General* to 801-93 *Analytic mechanics*. However, one of these subdivisions, 71-90 *Instruments and machines,* has a further subdivision 75-76.95 *Calculating machines* broken out. That subdivision has a subdivision 75.5-76.95 *Electronic computers. Computer science* broken out, which in its turn has a subdivision 76.75-76.765 *Computer software* identified at this level.

We are to interpret this schedule as indicating that each named subclass has very approximately the same number of books. Where a class has a distinguished subclass, the remainder of the class has the same specificity as the other classes. In this case, the size of the literature in *Computer software* is about the same as that in *Electronic computers. Computer science* excluding *Computer software, Calculating machines* excluding *Electronic computers, Computer science,* and *Instruments and machines* excluding *Electronic computers. Computer science.* All of these are about the same size as 440-699 *Geometry. Trigonometry. Topology*.

In this sort of system, "by about the same size" means differing by only one or two orders of magnitude – a factor of 10 to 100 difference in the number of books catalogued. Remember that the LOC system is designed to catalogue every book ever published and those expected to be published in the foreseeable future. In a total population of perhaps 100 million, a factor of 10 or even 100 difference in size is small.

The schedule for *Law* exhibits the same characteristics. It has up to four levels of hierarchy, with most of the more specific classes leaving room in the more general class for works not otherwise classified. See for example 670-709 *Domestic relations. Family law*.

Semantic coherence

Order in the LOC schedule is maintained by the sequence of the codes attached to the class names. At the highest level, we see that the classes proceed from general works, through philosophy, religion and psychology, through history and the social sciences, then law, the arts and literature, then the physical sciences, medicine and various technologies. Library science is at the end. This last is probably because the whole LOC indexing system is a product of library science, so that library science has a special status as a meta-discipline.

A much larger picture of the semantic relationships shown by the sequence of class names is given by the law sub-schedule. The sequence of class names at the same level is frequently intuitive, but at least generally not jarring.

Maintainability

Having been in use and in continuous development for more than 100 years, the LOC schedule shows much evidence of how it is maintained. New topics are added by breaking out new subdivisions of existing classes, as shown in an extreme way by the placement of *Computer software* at 76.75-76.765, which shows a long-term expansion of the literature related to computing from what was originally a fairly minor branch of mathematics: 71-90 *Instruments and machines*. The decimal numbering system allows indefinite expansion by inserting new codes between existing ones. Contraction is handled by letting declining literatures be absorbed back into the more general class.

New relationships between disciplines and interdisciplinary fields are dealt with by specializing one of the original areas, for example K366-380 *Sociology of law. Sociological jurisprudence* and K190-195 *Ethnological jurisprudence. Primitive law*. As to relationships between disciplines, note that the class K670-709 *Domestic relations Family law* has a corresponding class in *Social Sciences* HQ811-960.7 *Divorce* (not shown in Appendix D). The former is used for legal issues while the latter for more sociological issues, and both are subdivisions of more general classes (respectively *Law of persons* and *The family. Marriage. Home*).

Workability

By "simple subject" is meant a subject which can be described by a single term, such as *law* or *information retrieval*. The requirement that each simple subject in a domain has a clear place in the schedule of an indexing system for that domain is essentially a requirement for exhaustivity. Some simple subjects are very specific, so must appear in the specialization hierarchy.

Consider, for example, the programming language SQL. There is a classification QA76.73 *Programming languages individual languages*. That classification has a process for generating subclasses for specific languages, which yields the classification QA76.73.S67 for *SQL*. However, some works mainly about SQL are considered to fall under the classification *Databases*, for example *A Guide to the SQL Standard: A User'sGuide to the Standard Relational Language SQL* by C.J. Date (call number QA76.9.D3 D369 1987). Note that the class *SQL* is not a subclass of the class *Databases*, but rather of the class *Programming languages*.

A complex subject is one which requires a compound term to describe, such as *Sociology of law*. Not only must there be a place for it, as there is in the Schedule shown in Appendix D, but there must be clear guidelines to identify it as a subclass of *Law* rather than *Sociology*. This can be somewhat subtle. Consider the complex subject *Information technology in law*. The University of Queensland library has a number of books on that topic, many of which are classified under *Legal research*, for example *Technology for Law Offices* by S.K. Andrew (call number K87 L37 1997). However, *The Millennium Bug: Its Impact on the Legal and Accounting Professions* by Michael Blake et al. is classified under *Computer software* (call number QA76.76.S64 M552 1998).

The published schedule

Classification systems are applied and used by humans. Therefore, the schedule must be published. Because systems like the LOC system are so large and complex, control of the principle of uncertainty is a major factor in their design and management. For a start, people need to be trained in the use of the schedule (one element of the training of a library professional). However, even though the managers of the LOC system assume that the cataloguers employing it are well trained, it is necessary to include with the published schedule instructions as how to deal with ambiguous cases. For example, the class QA76.9.A93 *Auditing* in the schedule has the note attached

> Class here works on the auditing of electronic data processing systems and activities. For works dealing with the auditing of electronic data processing departments, see HF5548.3

An extreme example is the following note attached to the schedule entry TD *Environmental technology. Sanitary engineering*:

> The promotion and conservation of the public health, comfort and convenience by the control of the environment. Cf GF Human ecology HC68, HC95-710 Environmental policy (General) HD3840-4730, Economic aspects (Government ownership, municipal industries, finance, etc.) QH75-77 Landscape protection RA565-604 Public health S622-627 Soil conservation S900-972 Conservation of natural resources TC801-978 Reclamation of land TH6014-7975 Building and environmental engineering

Notice how diverse the classifications which must be considered before a work is finally assigned to a subclass within TD. A small difference in judgment can result in a work being assigned to widely differing classes.

The LOC schedule is so large and diverse that it is maintained in sections, as shown in Appendix E. There are 44 individual schedules, which are independently revised every so often. Of the 44, 31 have been revised in the 1990s, 10 in the 1980s, 2 in 1976 and one (PG *Russian literature*) most recently in 1948.

Finally, the LOC system is used by thousands of libraries. One way the LOC attempts to control the principle of uncertainty in that environment is to make use of its status as a legal deposit library – a copy of every book published in the United States must be lodged in the Library of Congress. Therefore, the LOC cataloguing staff must classify every book, and therefore books published in the United States have included in them a skeleton of a catalogue entry called "Library of Congress Cataloging-in-Publication Data" which includes a call number and also subject descriptors. In addition, it makes available its catalogue entries to other libraries. This very greatly reduces the number of times a particular book is assigned a call number.

The classification schedule is generally not made available to the user of a library, at least as such. Readers can search for books by author, title and subject descriptor, but not by the name of the LOC classification associated with its call number. Of course, books with similar call numbers are located near each other on the shelves, but the lack of availability of classification schedule enormously reduces the principle of uncertainty at a cost of a certain arbitrariness – the reader will find the book using the other descriptors, but may be surprised at where it is shelved.

Dewey Decimal System

We have considered the Library of Congress classification system in some detail not because we expect the reader to have become proficient in that system, but as an extended example of a very large and evolving classification system whose scale brings out significant issues in the design and engineering of classification systems. To cast further light on these issues, we will take a briefer look at another classification system used in many libraries, the Dewey Decimal Classification System (DDC). The DDC is of a similar scale to the LOC and has an even longer history.

The first observation is that even though the DDC covers the same domain as the LOC, it is quite different. Comparing the most general levels of the two systems (the DDC in Appendix F), we see that the DDC with a maximum of 10 subclasses for any class is much narrower than the LOC, which has 23 at the top level, and 24 for *general law*.

This difference in width somewhat obscures the fact that some things are classified very differently, reinforcing the point made in Chapter 10 that classification systems are essentially arbitrary, and therefore subject to design. Consider computing-related topics. In the LOC, computing is a subclass of QA *Mathematics*, 71-90 *Instruments and machines*, 75-76.95 *Calculating machines*; while in the DDC, computing-related topics are a subclass of 00 *Generalities/ Generalities*. (*Mathematics* is a subclass of major class 5 in the DDC.) In both cases, this represents a now very large literature which came into being very long after the overall classification system was established, and the consequence of different design choices as to which class to expand with the new literature.

Notations for the two systems are also very different. The LOC uses a mixture of letters and numbers, so has a large number of subclasses for any class. It was designed for very large libraries with a need for a large number of classes to retain the desired level of specificity. The LOC system is very widely used in university and research libraries. The DDC's strictly decimal notation has the great virtue that the class assigned to a particular work can easily be more or less specific depending on the size of the corresponding literature in a particular library. A library with only a few computing books can assign them all to the general code 005, while a large collection can subdivide to a further four or five levels. This makes the DDC more suitable for smaller libraries, and particular for systems of libraries which include both large and small, such as operated by many governments.

Note that this flexibility is not a property of the DDC classification system as distinguished from the LOC classification system. The systems of class names in either case can be deeper or shallower depending on the size of the collection and the degree of specificity required. It is a property of the system of notation – the DDC notation is more flexible than the LOC.

Types of classification schemes

Looking at these large and long-lived classification systems gives us an opportunity to revisit and extend the discussion from Chapter 8 on the design of classification systems. Both the LOC and DDC are large single hierarchies. The DDC system is ultimately based on a theoretical analysis, while the LOC system is based on literary warrant. Of course, the great expansion of the DDC system during its history has led to considerable distortion of its original structure.

These sorts of systems, built as a single hierarchy, are called *enumerative* systems. During the development of the system of classes, each class is subdivided separately, with the subclass names being developed as appropriate to the problem and the particular superclass (relative semantic dimensions). Thus in the LOC, K *Law* is subdivided into a large and heterogeneous number of subclasses, while QA *Mathematics* has nine subclasses, six of which are branches of mathematics and two are tools used in calculation (47-59 *Tables* and 71-90 *Instruments and machines*), with the final class 1-43 *General* needed for exhaustivity.

However, the DDC system has a more regular method of generating subclasses. There is a system of tables. Table 1 *Standard Subdivisions* includes subclasses for philosophy and theory, dictionaries etc., serial publications and historical and geographical treatment. Table 2 *Areas* is a hierarchical set of names of regions, countries, and political subdivisions of countries. There are five other tables.

Using these tables, a dictionary is represented by the code 03 from Table 1, a serial publication by 05 from Table 1, and the USSR (now dissolved) by the code 47 from Table 2. In this system, a dictionary of computer science would have the code 004.03, while a dictionary of international law would have the code 341.03. A journal of computer science in the USSR would have the code 004.0547 (the tables are applied in sequence). This regular method for generating subclasses simplifies the application of the system, and makes the schedules smaller.

Tables simplify the DDC to a degree, but play a small role in the system as a whole. It is possible to elevate this use of multiple independent absolute semantic dimensions into a method of designing large schedules. We have already seen an example of this approach. The hierarchy of classifications for noodle restaurant dishes in Figure 8.4 is derived from an arbitrary sequence of choice among the three dimensions of Figure 8.1. A different sequence would result in a different hierarchy.

This new method of building a classification hierarchy by combining several different absolute classification systems is called the method of *facets* (each independent absolute system is called a facet.). *Faceted* systems have been developed in response to a number of problems with large enumerative systems.

A large system is expensive to maintain. It is particularly expensive to make large systematic changes. To take a hypothetical example, the USSR was the subject of a large body of literature, some of which was historical material concerning its constituent parts such as Byelorussia, the Ukraine and Uzbekistan. With the dissolution of the USSR in 1991, one would expect that separate classifications would be established for new material relating to each of these now independent countries. However, the historical material for each of these countries would have to be re-classified. Furthermore, the relative importance of various subdivisions would change – *Cossacks* might appear as a major subclass in the hierarchy relating to the Ukraine, while literature about the silk road might be a major subclass in the hierarchy relating to Uzbekistan. Further, these two classes are likely to be much more important in the new system than they might have been in the system relating to the USSR.

Complex subjects are problematic in enumerative systems, as we have seen above. When classifying material about information technology as used in law, we have to decide whether the focus is primarily on information technology or primarily on law. This means that a coherent body of literature can be split into two very distant classifications. Furthermore, giving a named place to a complex topic can result in excessive specificity in the classification system, since it requires subdivision of one of the simple topics which may already be as specific as the design allows.

For these two reasons, large enumerative systems tend to be rigid.

Before proceeding to a consideration of the faceted alternative building classification systems, it is worth looking at a technical consideration. Large enumerative systems like LOC and DDC were developed at a time when the only way to publish the schedule was in print. These systems have many hundreds of thousands of most specific classes, so that the published schedule runs to many thousands of pages. Even though a system such as used in Figure 8.4 is flexible, in that there could be several different hierarchies, if the system is large and must be published in print, the costs involved and the physical size of the printed schedule make it very difficult to publish more than one sequence.

Today, on the other hand, it is possible to store the schedule in a computer system. A user can navigate through the schedule by choosing one class from each semantic dimension in any convenient sequence, so that there can be many ways of reaching a particular most specific class. To return to the noodle restaurant example, the menu in the restaurant is in fact presented as in Figure 8.1 rather than as in Figure 8.4. The diner is presented with three semantic dimensions of choice and can refine a selection in any sequence desired.

There are many historical examples of faceted classification systems. Most statistical tables have at least two, sometimes three or four facets. An outstanding example is the Periodic Table of the Elements, one of the foundations of modern chemistry and nuclear physics, which has essentially two semantic dimensions, *valency* and *atomic number*. However, in the pre-computer age faceted systems were nearly all very small.

A large contemporary example of a faceted system is the SNOMED system for classifying medical treatment incidents. As shown in Table 11.1, the SNOMED system has more than 150,000 class names organized into 11 facets. A particular incident, say a fever caused by an infection in the throat by a particular bacterium treated by a particular antibiotic, would be classified using names from *Anatomy*, *Living organisms*, *Symptoms*, and *Chemicals and drugs*, at least. There are therefore a vast number of possible classifications.

We can observe several things about the SNOMED system. First, all of the facets are large. The smallest, *Social context* has more than 1000 class names, while the largest, *Diagnoses*, has more than 40,000. These facets are therefore all enumerative

Table 11.1 The facets of the SNOMED system.

Topography (Anatomy)	13,165
Morphology	5,898
Diagnoses	41,494
Procedures	30,796
Functions, symptoms, etc.	19,355
Living organisms	24,821
Chemicals and drugs	14,859
Physical agents, forces	1,601
Social context	1,070
Occupations	1,949
General modifiers	1,594
Total	156,602

systems in themselves – much smaller than the LOC or the MC, but comparable in size to the SIC codes from Appendix A.

Secondly, the facets are generally not exhaustive, in that most medical incident reports would be described by only some of the facets. If no drug is prescribed, then no classification from *Chemicals and drugs* would be used. If a person has a systemic disease like chicken pox then no classification from *Topography (Anatomy)* would be used. Of course it is possible to expand each of the facets with a class name *Not applicable*, which would then make each of them exhaustive.

Thirdly, the entire classification system is never used in practical analysis of the data collected. The total number of classifications possible, if all eleven of the facets are used in every incident report, is more than 10^{40}. If every person on earth had 10 incident reports per year for their entire lives, there would be a total of fewer than 10^{13} incident reports after 100 years. This is to say that even if each incident report had a different classification, only one in 10^{21} classifications would have an entry. This is equivalent to 1/10 of a microsecond during the entire history of the earth. (The point here is partly that, due to combinatorial explosion, faceted classification systems can generate enormous numbers of classifications.)

The SNOMED system is used for transaction purposes – insurance reimbursement funding categories are often based on high levels of the code system, and the system provides a controlled vocabulary for keeping medical records. The SNOMED system also supports data warehousing type queries (such as described in Chapter 9) on large populations of incident reports. Each query involves only a few of the facets.

Finally, we observe that the degree of specificity of classification differs greatly between incident reports. For example, a routine antibiotic treatment for a sore throat would be coded in the *Topography/Anatomy* facet with a fairly general category indicating the throat, while a treatment for a small tumour would be coded with a much more specific class denoting the specific structure in the throat on which the tumour was found. In the routine sore throat incident, the report would use a general class from the *Living organism* facet indicating unspecified bacterium, while an unusual case would require a laboratory analysis of the organism and a much more specific class from that facet.

This variation in specificity of reports has significant technical implications for both information retrieval queries and data warehousing type aggregate queries, since the data may be coded more or less specifically than the terms used in the query. If the query terms included the code for the upper part of the larynx, the person making the query would have to consider what to do with the routine sore throat reports which are coded much more generally. Similarly, if the query term were for an unspecified bacterial infection, the user would have to decide what to do with reports coded for specific bacteria. These issues are covered in the data warehousing literature, and are outside the scope of this text.

So far we have looked at enumerative and faceted methods as alternatives for the design of classification systems. There is in practice much scope for using both methods. We have already seen with the SNOMED example that facets can be so large that they are hierarchical enumerative systems in their own right. The SNOMED example has also shown that facets need not be exhaustive. This last will now be developed into the use of facets as relative semantic dimensions rather than absolute.

We have already encountered this phenomenon in Chapter 9. The classification system for noodle restaurants in Figure 8.1 indeed includes a faceted system with three facets, but this faceted system applies to only one of the two most general classes in the menu, *dry noodle dishes*. The other most general class, *wet dishes*, has only three subclasses, none of which are related to the faceted system by which the other general class is subdivided.

We might be classifying items appearing in a newspaper. The two most general classes might be *editorial matter* and *advertisement*. Editorial material might be classified with two facets:

1. Field, consisting of general news, sports, business, higher education, information technology, etc.
2. Genre, consisting of news item, feature, opinion, letter, etc.

Advertisements might be classified with two different facets: the first being *classified* or *display*; and the second being the SIC code from Appendix A.

None of these four facets is exhaustive, but the first two are exhaustive relative to the most general class editorial matter, while the last two are exhaustive relative to the most general class advertisement.

Such systems of *relative facets* (often called *differential facets*) are widely used.

Developing faceted classification systems

Chapter 8 has covered in some detail the methods of developing enumerative classification systems. These methods, theoretical analysis and literary warrant, scale up to the size of systems as large as the LOC or DDC. The largest system of all, the Linnean nomenclature for living beings, is based on a strict theoretical analysis. Furthermore, Chapter 9 considered a theoretical view of designing what was called there "systems of absolute semantic dimensions", and which we now see as faceted systems. This section therefore concentrates on a literary warrant method for developing faceted systems.

First, we need some terminology. A *faceted system* consists of several *facets*. A facet has a *name*, and is composed of a set of class names called *foci* (singular *focus*). Facets are sometimes called *axes* (the SNOMED system) or *modules*. The design problem is how to come up with class names, facet names, and how to group the class names into foci of particular facets.

A literary warrant approach starts with a body of objects to be classified. The process has several steps:

1. Since the classification system is composed of names, we first create a title-like description of each object.
2. We examine the sample of object titles to identify candidate terms called *isolates*.
3. Similar isolates are grouped together, and the groups given names, some of which may themselves come from the set of isolates.
4. Isolates are evaluated for specificity. Too-specific isolates are replaced by suitable more general terms.
5. We then evaluate the named groups of isolates as potential semantic dimensions for classification of the objects. Those groups deemed suitable are now facets, and their constituent isolates become foci.
6. We make sure the new facets are complete by seeing if the foci present suggest others, make hierarchical organizations where necessary, and generally clarify the foci, before finally declaring the new faceted system complete.

For example, consider a collection of a few hundred newspaper items relating to information science as presented in this text. Summaries of three samples from this collection are presented in Figure 11.1. The first task is to create title-like descriptions for each item.

It should be clear that the headlines are not suitable. Rather than describing the news item, they are designed to call attention to it on the page, so do not yield reliable isolates. We need single-sentence descriptions of the content, such as:

A. Publishers of scientific and scholarly journals are linking up on the Internet to make it easier for scientists to do research.

B. Search engines are increasingly using ranking engines, natural language processing, multimedia and metasearchers.

C. C&W Optus has a wireless application protocol (WAP) mobile phone service, as will Telstra and Vodafone.

The first of these came from the first sentence of the article. The second came from the last sentence of the summary, together with a replacement of "Many traditional players" by "Search engines". The third required composing a new sentence using

A. Science link online

TWELVE publishers of scientific and scholarly journals are linking up on the Internet to make it easier for scientists to do research. The agreement allies some of the biggest rivals in the lucrative arena of scientific publishing, including Oxford University Press, Macmillan Magazines and Elsevier Science. They say the agreement will link three million articles at first, and even more later.

B. Searching for a smarter engine

FOR most Internet users, making an inquiry on a popular search engine is akin to rummaging for the proverbial needle in a haystack. Literally millions of matches – the vast majority meaningless – flood back from a keyword search. The search capabilities offered by the larger sites can take too much time and effort for the average user to persist. These days, improving relevancy to users calls for smarter search engines, not just bigger ones, analysts say. While the traffic on sites such as Alta Vista, Yahoo! and Excite continues to grow, search engine usage patterns are fragmenting. Many traditional players are increasingly using ranking engines, natural language processing, multimedia and metasearchers.

C. WAP on line for Net big thing

JAMES Packer has caught on to it. Cable & Wireless Optus pipped its competitors to launch it. And every media company has begun to develop applications for it. It's WAP, wireless application protocol, the new generation mobile phone that is more than just a mobile phone. Mr Packer, chairman of Publishing & Broadcasting, which owns the Nine Network and 80 per cent of ecorp, calls it the portable Internetconnected device. Mobile phone company Nokia chief executive Jorma Ollila calls it Internet in the pocket. Next week, Australians will be able to get their first taste of WAP when C&W Optus launches its service with the Nokia 7110 handset. Telstra and Vodafone are also planning WAP services.

Figure 11.1 Information science-related news stories.

fragments of several from the article summary. We see that creation of these content descriptions is a difficult task, to a degree similar to the abstracting problem discussed in Chapter 7. This process has been applied to a larger sample in Figure 11.2.

The next step is to identify isolates by selecting terms from the description statements which best represent the contents. This certainly excludes stop words like "the" or "with" and generic words such as "give" or "shows". Figure 11.3 shows the result of this process on the description statements in Figure 11.2.

Now we have a first draft of isolates. Some of these may be too specific for classification purposes. Set B includes a number of specific advanced search technologies, while set C includes a number of specific telecommunications companies (telcos). In the set of refined isolates in Figure 11.4, the specific advanced search technologies have been replaced by the more general term "advanced search technology", and the specific telcos have been replaced by the more general term "telcos". In D the term "customers" has been replaced by the more instrumental term "end users", while in G the term "unreliable" has been replaced with "reliability" and in J "maintainable" by "maintainability". In J also, the specific term "Victoria" (name of an Australian state) has been replaced by the more general term "Australia".

In K, a more general but more focused "embedded technology supplier" has replaced the specific term "General Motors". In L and N "people" have been replaced by "end users" for the same reason as in D. In N, the collection of isolates "control",

A Publishers of scientific and scholarly journals are linking up on the Internet to make it easier for scientists to do research.

B Search engines are increasingly using ranking engines, natural language processing, multimedia and metasearchers.

C C&W Optus has a wireless application protocol (WAP) mobile phone service, as will Telstra and Vodafone.

D Customers must control Internet marketing data.

E Data warehouse of marketing data established in Australia.

F Conversion of Web sites to mobile markup languages.

G Search engines can be unreliable.

H World atlas available on the Internet.

I Web site developers must understand the small businesses which provide content.

J Victorian election result shows that Web sites must be maintainable.

K Voice-activated access to the Internet in General Motors cars.

L Web sites enable people to store photographs online.

M Internet bandwidth doubles every six or nine months.

N The Internet will give people control over how they interact with the economy.

O Internet searches can be launched with words from any application.

Figure 11.2 Sample content descriptors.

A   Journal publishers, linking, Internet, scientists, research.

B   Search engines, ranking engines, natural language processing, multimedia, metasearchers.

C   C&W Optus, wireless application protocol (WAP), mobile phone, service, Telstra, Vodafone.

D   Customers, control, Internet, marketing data.

E   Data warehouse, marketing data, Australia.

F   Web sites, mobile, markup languages.

G   Search engines, unreliable.

H   World atlas, Internet.

I   Web site developers, understand, small businesses, content.

J   Victorian, election result, Web sites, maintainable.

K   Voice, Internet, General Motors, cars.

L   Web sites, people, photographs.

M   bandwidth.

N•   Internet, people, control, interact, the economy.

O   Searches, any application.

Figure 11.3 First cut isolates.

"interact", "the economy" has been replaced by the more specific and compact "retail economy". Finally, in O, the very general term "any application" has been discarded.

A   journal publishers, linking, Internet, scientists, research

B   search engines, advanced search technology

C   telcos, wireless application protocol (wap), mobile phone, service

D   end users, control, Internet, marketing data

E   data warehouse, marketing data, Australia

F   Web sites, mobile, markup languages

G   search engines, reliability

H   world atlas, Internet

I   Web site developers, requirements, content providers

J   Australia, election result, Web sites, maintainability

K   voice, Internet, imbedded technology supplier, cars

L   Web sites, end users, photographs

M   bandwidth

N   Internet , end users, retail economy

O   searches

Figure 11.4 Refined isolates.

Having refined the isolates, we now organize them into a first draft of facets and foci in Figure 11.5. The terms appearing as foci are all derived from the set of isolates (each term carries with it the identifiers of the documents in Figure 11.4 from which it is derived). Each facet has been given a name, which in some cases also is an isolate from Figure 11.4. Note that the isolates "service" from C and "control" from D have been discarded (the last facet "not used"), on the grounds that they do not appear to be useful as classifiers.

In the last stage, we refine and organize the facets, as shown in Figure 11.6. Several additional foci have been added (those without a document identifier) on theoretical grounds. For example, if we have the development process and the stage *requirements*, then we need the other standard stages. In some cases such as in the facet *content providers* we expect that we will find other classes than *journal publishers*. We have also introduced some structure, so that *players* and *infrastructure* are both hierarchies. We have removed some foci – "data warehouse" from *Content* in favour of the more specific and operational "marketing data" and "Internet" from *Medium* because it is too general. We have also removed one facet *Nature of content* because it did not seem to lend itself to a definite exhaustive set of foci.

The final set of facets in Figure 11.6 is now quite a reasonable classification system. It is plausible that there would be news items about:

- the *construction: Development process*
- by a *telco: Infrastructure provider: Player*
- of *advanced search technology: Search engines: Infrastructure*
- where the concern is *reliability: Quality*
- of *voice: Interaction mode*
- over *mobile phone: Hardware*
- to *web sites: Medium*

Development process: requirements I
Content providers I: journal publishers A
Infrastructure providers: telcos C, Web site developers I, imbedded technology supplier K
End users D L N: scientists A
Infrastructure: search engines B G, advanced search technology B, markup languages F, wireless
     application protocol (WAP) C
Quality: reliability G, maintainability J, bandwidth M
Interaction mode: voice K
Hardware: mobile phone C F, cars K
Medium: Internet A D H K N, Web sites F J L
Nature of content: linking A
Nature of use: research A, searches O
Content: marketing data D E, data warehouse E, world atlas H, election result J, photographs L
Context: retail economy N
Not used: service C, control D

Figure 11.5 First cut facets.

- for *overview: Nature of use*
- of *election results: Content*
- in *Asia: Location*
- where the concern is *equity: Context.*

Such an article would be described by one term from each facet. Of course, like SNOMED, none of the facets is exhaustive, so that only a few facets would classify a typical article. Although none of the individual facets is exhaustive, the system as a whole is.

We emphasize that most of the steps require considerable design input. The most difficult design step, requiring the most judgment, is the assignment of title-like content descriptors. This step is also the most important, as the success of the rest depends on it. Creating the isolates is fairly mechanical in that a list of stop words can be created, but there is some judgment in the example shown. Refining the isolates requires significant design input, where the nature, purpose and size of the collection are relevant. Judgment is required in the organization into facets, and also in the facet refinement process. Mechanical help is useful in managing the identifiers, computing occurrence frequencies, and other statistical processes which can inform the design decisions.

This literary warrant approach to development of a faceted classification system is quite powerful. Taking advantage of a body of material to be classified, it allows the designers to combine their knowledge of the domain and purposes of the system with a view of its actual content, and can often produce a superior classification system.

Development process: requirements I, construction, evaluation, maintenance
Players:
    Content providers I: journal publishers A, other classes of content providers
    Infrastructure providers: telcos C, Web site developers I, imbedded technology supplier K, database providers, etc.
    End users D L N: scientists A, various other distinguished classes of end user
Infrastructure:
    Search engines B G, standard technology, advanced search technology B, policy, coverage etc.
    Markup languages F, HTML, XML, wireless application protocol (WAP) C, Others
    Other technologies, with subdivisions
Quality: reliability G, maintainability J, bandwidth M, currency, validity
Interaction mode: voice K, GUI, browser, virtual reality, etc.
Hardware: mobile phone C F, cars K, household devices, workstations, etc.
Medium: (removed: Internet A D H K N), Web sites F J L, CD-ROM, databases (Facet removed) Nature of content: linking A,
Nature of use: research A, searches O, overview, visualization, etc.
Content: marketing data D E, data warehouse E (removed), world atlas H, election result J, photographs L, other distinguished classes of content
Context: retail economy N, business-to-consumer, business-to-business, social impact, equity, etc.

Figure 11.6 Final facets.

Notation

Now we have a classification system. It may be enumerative or faceted, and it may have been developed using theoretical analysis, literary warrant or some combination. We now must consider how it is represented, which means that we must think about what it is to be used for, and in what circumstances.

Locating objects in physical space

In library-type applications, the classification system is generally used to locate the objects in physical space. This is true whether the objects are books, videotapes or museum specimens. The arrangement in space is intended to reflect the semantic relationships among the objects – similar objects are near each other. The spatial arrangement is often linear – a collection of shelves traversed from left to right top to bottom, but sometimes in more dimensions – say, display cabinets in rooms or specimen drawers.

In order for the classification system to reflect the physical arrangement, it must be possible to arrange the class names in the appropriate sequence. If the system is enumerative, then a simple way to get a linear sequence of most specific class names is to sequence all the subclasses of each class, so that the first most specific class is the first subclass of the first subclass, and so on. All the examples of enumerative systems in the appendices are sequenced in that way.

If we have an exhaustive multifaceted system, we can achieve a sequence by designating a sequence for the facets. If we accept the sequence of the set of facets in Figure 11.6, then the different values of *Concern* will be adjacent, and the different values of *Development process* furthest apart. If the facets are not exhaustive (as in fact the facets of Figure 11.6 are not), as we have seen above they can be made so by adding the class name "not applicable" to each facet. This is in effect representing the faceted system as a hierarchy in the same spirit as the way Figure 8.4 represented the restaurant menu of Figure 8.1 as a hierarchy.

Now that we have a sequence, the last requirement is to represent it in some sort of notation. The usual approach is to mirror the class names with a system of symbols which has a conventional interpretation as a sequence – most commonly a decimal system of some sort, like the DDC (Appendix F) or the SIC (Appendix A), or an alphanumeric system with more than ten subdivisions such as the LOC system (Appendix D). Most wordprocessing systems have a number of ways of recording hierarchical systems of terms, providing further examples.

A final use for the notation is to label the objects so that the object itself indicates where it is stored. Library books generally have their call numbers written on the spine, for example.

Locating objects in information space

Many populations of objects are not physical, but are stored, retrieved and presented by a computer system, or perhaps the objects are stored in file cabinets by some kind of file accession number which is returned by the computer system and employed by the user to fetch the actual object.

In this case, the enumerative system can be represented as a menu which can be expanded as successively more specific choices are made. There is still a sequence implied, which is the same sequence as in the previous section. The faceted system is quite different in this environment, however. The various facets can be presented to the user as a computer form, and the user can select foci in any sequence. The programs managing the form can construct the identifier of the class from the facets selected, and the catalogue entries returned in an appropriate sequence. In this case, the notation used can be invisible to the user, who can work directly with class names.

Even in fully computerized systems, however, it is very common for each object's catalogue entry to have its classification represented in some notation. The SNOMED system described above is used for medical incident reports which originate in medical offices or hospitals and which must be transmitted to, for example, health insurance companies. It is very common to use a compact notation for this purpose, as it makes computer processing somewhat easier.

For enumerative systems, the decimal code can be used. For multifaceted systems, it is usual to represent the facet name as well as the focus applied to the object, as many facets as are required for the object. One way to do this is to encode the facet names as XML markup elements and the foci as permissible content, as described in Chapter 7.

Key concepts

An **index** is a systematic guide to the contents of an information space. An **indexing system** is a procedure for applying an **indexing language** to a particular collection. The indexing system includes a **schedule** (the descriptor terms) and a **notation** (a way to represent descriptors in a condensed form.).

There are two basic kinds of indexing systems used in classification: **enumerative** (based on a single hierarchy) and **faceted** (based on several semantic dimensions). In a faceted system each semantic dimension is called a **facet**. Each facet is a possibly hierarchical collection of descriptors called **foci** (singular **focus**).

Further reading

Most textbooks in library science include chapters on the major classification systems, and there are books on specific systems. A more technical view of classification may be found in Salton (1989: Ch. 9).

Formative exercise

Find a large classification system in some domain which you understand.

●   Is it enumerative, faceted, or both?
●   When was it first designed?
● Does it have computer-based access methods?
● Does it appear to be easy to add a new simple or complex class name to the schedule?

- Who makes use of the index?
- Does the index have instructions to the classifier? How complex are they?
- Does it have a notation? What is it used for? How is it organized?

Tutorial exercises

Develop a faceted classification system for television broadcast units (Chapter 8). Use the classifications you developed for the tutorial exercise there. A possible set of classifications includes:

Australian outback shows
Movies
Wildlife shows
Comedy sketch shows
Music
Ad for product
Amateur video shows
Station logos
Ad for service
Foreign language programmes
Public service announcements
Horror
Situation comedies
Morning
News
Entertainment
Midday
Current affairs

Hospital/medical shows
Ad demonstrating something
Real life shows
Soap operas
Sitcom – work based
Game shows
Police shows
Sitcom – family based
Variety shows
Westerns
Sitcom – friends based
"company" shows
Ads
Sitcom – place based
Talk shows
Genre
Overnight
Documentaries
Audience
Prime time

Promos
Series
Ad reminder
Test patterns
Mini-series
Ad group programmes
Historical documentaries
G, PG, MA, M (audience suitability ratings)
Sport
Cultural documentaries
Saturday morning
Children's programmes
Teleshopping
Lifestyle
Timeslot
Cartoons
Drama

Open discussion question

An ideal faceted classification system has a slot for each selection of one focus from each facet. In practice, it often happens that some combinations do not make sense. How does a faceted classification system with many combinations not making sense differ from an enumerative system? Or does it make sense to think of the two methods as different ways of looking at the same problem?

12 Descriptors

In Chapter 4, we saw that an object is represented in an information space by a surrogate which contains descriptions of its content, and we looked at some issues in descriptor design. Content descriptors are similar to classifiers, and in Chapter 11 we looked at methods for constructing large classification systems. This chapter examines the related engineering issues in the design of large systems of descriptors.

Introduction

A descriptor is a representation of the content of an information object. A classifier is also a representation of content, but as we have seen in the last few chapters, classifiers have several other functions as well. As a consequence, an information object can have only one classifier. The photograph of Mrs Prudence Morgan milking a Guernsey cow in Iowa, around the turn of the 19th century, which was discussed in Chapter 4, is a good example. One could imagine that the photo is classified as *Rural America 1875–99*, if it were in a collection of 100,000 photographs from all over the world covering the period 1860 to the present. If it were in a collection of industrial photographs, it might be classed as *Dairying – pre-industrial*. In either case, the classifier provides a very general descriptor.

However, as we saw in Chapter 4, a user might have a much more specific information need. Furthermore, at the level of specificity of a query, the photograph is a complex object. The most specific descriptors might include "Prudence Morgan", "milking", "Guernsey cow", "rural Iowa", "late 19th century". If Prudence Morgan is not a famous person, then the less specific descriptor "rural housewife" might be preferred. So the classifier is not a satisfactory descriptor. It is either too general *(Rural America 1875–99)*, or only partial *(Dairying – pre-industrial)*.

Collections of information objects therefore usually have both classifiers and descriptors. Just as large collections have design problems for classifiers due to scale, so also do large collections have design problems for descriptors.

There are two basic strategies for descriptor design, the *subject list* and the *thesaurus*. These two strategies roughly correspond to the enumerative and faceted strategies for developing classification systems, which we have just seen.

Design requirements

We first look at the tasks we expect our sets of descriptors to perform, so we can understand the constraints governing their design.

The main purpose of a set of descriptors is to reduce the variety in the surrogates of the information objects. We saw in Chapter 4 that we need to have enough variety to distinguish the information objects. The problem is that in most collections there is generally much more variety than is necessary. The additional variety comes from the fact that generally many different people are involved in the preparation of the information object over a considerable period of time. Variety comes from:

1. The syntactical rules of the language in which the object is described. The text strings "computer", "computers", "computing" are all used to describe the same concept in different linguistic contexts. Remember from Chapter 3 that the ISAR system works by matching text strings, not by understanding the text. Differences between English and American spellings are a similar source of variety.

2. Different ways of expressing the same concept. A writer might use "computing" in one context, "computation" in another, and "calculation" in a third. These refer to somewhat different ideas, which could easily be the same from the point of view of someone making a query.

3. Evolution of a terminology over time. In 1999 what are generally called "vertical portals" were introduced. At that time they were often called "specialized portals" or "limited domain portals". The term "vortal" was used by some people, but it did not win wide acceptance.

4. Historical differences. In established fields there are often differences in terminology arising from a number of different groups who have contributed to the field's development. One sees the terms "information retrieval", "text retrieval", "information storage and retrieval", "ISAR", and "database" all used to describe what to a person making a query are the same concept.

These sources of variety can be thought of as arising from a generalization of the principle of uncertainty.

The central task of a set of descriptors is to reduce variety by representing each concept in a preferred way. Systems generally have indexes which will direct enquirers to the preferred descriptor if they are using an alternate descriptor.

A second major task of a set of descriptors is to permit the user to browse the collection. Looking at what descriptors are used in a collection is a good way to get an overview of its contents. Since as we saw in Chapter 8 there is a limit to the number of different concepts a human can comprehend at once, the requirement to support browsing generally means that the descriptors are organized into a hierarchy. It is simple to annotate the descriptors, with the number of objects having that descriptor, giving a good summary of the contents of the collection. This is essentially the problem of content analysis as described in Chapter 4.

From functional requirements we turn to engineering principles discussed in Chapter 4. Our set of descriptors must be exhaustive – every object must be described by at least one descriptor, else it may as well not be in the collection.

We must also consider specificity; that is, we must decide how many objects are to be described by the average most specific descriptor. This is generally determined by the number of objects a user will tolerate in the response to a query and by the number of descriptors we expect a user to employ in a single query. If we have a collection of 100,000 objects, want the user to find objects with a single descriptor, and want the most specific descriptors to describe 10 objects on the average, then 10,000 descriptors are needed.

If we expect our users to employ two descriptors to make a query, then fewer descriptors are required, and they are less specific. If the descriptors are independent of each other, then our collection of 100,000 objects would need about 100 descriptors to retrieve 10 documents, since 100 ? 100 = 10,000 giving an average of 10 documents described by each pair of descriptors. In practice, descriptors are generally highly correlated, so it is more likely that 500–1000 descriptors would be needed.

Finally, we must decide on the width of the hierarchy to find out how many more general descriptors we need and how general they must be. If we permit our hierarchy a width of 20 – that is, each more general term has 20 more specific descriptors at the next level – then a collection of 10,000 most specific descriptors, would have two more general levels. There would be about 25 most general descriptors and 500 at the intermediate level.

We will now see how these design requirements shape some existing systems.

Subject lists

There are many collections of subject descriptors used in various applications. Some are highly specialized, such as those supporting the Medline medical research article database operated by the National Library of Medicine in the USA. Others are general collections intended to be adapted to specific purposes. The latter are often found in the library community, and often have long histories.

One such list is Sears' List, which is intended to be used by small libraries as the basis of their subject catalogues, and is in wide use. It was first published in 1923, and the 17th edition was published in 2000. It has about 6000 subject descriptors. It is designed so that one descriptor is sufficient to find a book. Since its scope is so broad, the most specific descriptors are semantically fairly general. A small library might have only 10,000 books covering just about any topic, and there would probably be several books described by any given subject descriptor. The list is not hierarchical.

Appendix G contains a selection from Sears' List, 1986 edition. The first selection is centred on the term *information science,* and the second on the term *computer software.* You can see that the terms are fairly general. There are only five terms related to the same field as information science, while although there are several terms related to computers, there are only five terms related to computer software.

A second example, which has a similar method of organization, is the List of Subject Headings published by the US Library of Congress. This list is widely used by large libraries and has a similar scope to Sears' List. It has about one million entries covering about the same range of topics as Sears' list, so it is much more specific. Appendix H contains the terms surrounding *information science* and *computer software.* You can see that there are a large number of terms which are not only related to *information science* and *computer software,* but are more specific whereas these terms are the most specific in Sears' List.

Neither list is strongly hierarchical. Sears' list is flat, while the LOC terms are hierarchically subdivided but only at a very specific level, so that the hierarchy is very broad. In Appendix H, the library records the number of volumes in the collection indexed by each subject term, but does not even make use of the hierarchy from which the most specific terms are constructed. Therefore neither system is very useful for content analysis.

A third example, the Association for Computing Machinery (ACM) keyword list in Appendix I is much smaller, having about 1500 terms. It has a much narrower scope, so is very specific. The ACM is a professional association in the computer industry, which publishes several journals and sponsors many conferences. Its descriptors cover a limited range of topics, but are much more specific than Sears' List. However, the LOC list is even more specific.

The term "information science" is a most specific descriptor in Sears' List. It has 35 more specific terms in the LOC list. The closest term to it in the ACM list is the more specific "Information Storage and Retrieval Systems", which has 39 more specific descriptors. The LOC has more than 240 descriptors more specific than "information storage and retrieval systems", so is a much more specific system than the ACM list.

In contrast to the other two examples which are flat, the ACM list is hierarchically constructed. The most specific descriptors are intended to be used together with their position in the tree, so the descriptor "Relevance feedback" would not be used on its own, but as "H.3.3 Relevance feedback". In this way the same term can appear in different contexts. For example, the term "virtual memory" appears both under Hardware (B.3.2 *Design Styles*) and Software (D.4.2 *Storage Management*).

Flat systems must also have a way of disambiguating terms. They generally do this by qualifying the term by another term in parentheses giving the context meant. For example, in Sears' List, we see "Inflation (Finance)", which distinguishes that use of the term "inflation" from its possible use in cosmology, say. The LOC typically has much more context included in the descriptor phrases. The two senses of inflation are represented there as "Inflation Accounting – see Accounting, Effect of Inflation On" and "Inflation Cosmological – see Universe".

The subject list reduces variety by prescribing a preferred description for every concept. In addition, the system must cater for the principle of uncertainty, that different people will describe the same concept differently. The main way of dealing with this is by extensive cross-references. The designer identifies alternative descriptions for concepts and includes them in the subject list, with a cross-reference to the preferred descriptor. Both Sears' and the LOC list have many cross-reference entries.

The ACM list, being much more specific and much more recent, does not. Its terminology has co-evolved with the computer industry. It does, however, have a cross-reference system to help the user find the most appropriate term. The choice of entry for "virtual memory" depends upon whether the focus of the article being described is on the hardware architecture or on programming techniques depending on virtual memory.

Subject lists solve the problem of descriptors by providing a preferred descriptor for each concept occurring in the collection. This leads to two problems. First, the descriptors often have to be extremely specific in order to be at the level of generality of the user making a query. Note in Appendix H the large number of LOC descriptors having only a single volume in a large university library.

Second, there is variety at the level of the syntax of constructing descriptors that is difficult to control. Suppose one were looking for information on the rebuilding of Paris under Napoleon III. A book containing information on the subject might have the LOC descriptor "France Literature and History Paris History 19th Century". How would one know how to construct the descriptor with that sequence of words? Suppose one were to look for a similar topic on Berlin in the 19th century. Would one think to look under "German Literature 19th Century History and Criticism"? Different groups of people develop different ways of describing things which might seem similar to someone looking for information. Further, different sections of large systems like the LOC are maintained by different groups of people. Maintaining overall consistency is very difficult.

Subject lists attempt to create single descriptors for topics for sound technological reasons. When these systems were designed, the main method for a user to access a library's subject catalogue was by a card file. A library with one million volumes might have 500 drawers of subject cards. It would be very difficult for a user to look up several different terms and combine the results. Being able to look up a single descriptor is much more practical.

Subject lists are often coordinated with enumerative classification systems. The LOC subject list is related to the LOC classification system, and Sears' List is coordinated with the Dewey Decimal System.

Thesauri

The introduction of information technology means that the user's interface is a computer screen rather than bay after bay of drawers. It is feasible to enter a more general descriptor which might have thousands of hits, and to refine the result set using other relatively general descriptors. For example, to find material on the rebuilding of Paris, one might start with "Paris", then add "history", then add "19th century". Alternatively, one might start with "history", then "19th century", then "Paris". The computer interface will produce the same result no matter what sequence the terms are entered.

Because of this much more flexible method of access, most more recently created collections of descriptors are collections of relatively general terms called, individually, *thesaurus* and collectively, *thesauri*. In its most recent editions, Sears' list has been changed from a list of subject descriptors to a thesaurus, as can be seen in Appendix J.

The most obvious difference from the earlier version in Appendix G is that the structure is no longer flat. For example, the term "information science" has below it "BT communications". This signifies that "communications" is a *broader term* (BT) than "information science", which is to say that a search on the descriptor "communications" would return catalogue entries described with "information science" (given appropriate software). Further, we see the entries "NT Documentation", "Electronic data processing" and so on. These terms are *narrower terms* (NT) than "information science". The term "information systems" is a narrower term than "information science", so in its entry, "information science" is shown as a broader term. It is typical for a thesaurus to have this broader term/narrower term hierarchy.

Additional structure can be seen under the entry "Information services". The entry "UF Clearinghouses, etc." shows that if one were looking for information on clearinghouses, for example, one should use the preferred term "information services" (UF stands for "use for"). This relationship is shown by the annotation "x" in the subject descriptor form in Appendix G. Where these terms appear in the list they indicate "USE information services". Note the entries for "Computer programming languages", "Computer programming" and "Computer programs".

Also under "information services" are entries labeled "SA" (see also), which point to other possible terms in the same way as the earlier version, and "RT Documentation, etc." which are *related terms*. The "see also" entries are relatively general terms which give the user hints as to other avenues for searching, while the "related terms" entries are simply other terms at about the same level of generality one might try.

Notice that there is some redundancy in the broader term/narrower term hierarchy. The term "information systems" occurs as a narrower term under both "information science" and "information services", but "information services" is itself a narrower term under "information science".

If you look at the extracts in Appendices G and J, the terms in the thesaurus version of Sears' list are essentially the same as in the subject list version. What has been added is the structural relationships. These structural relationships are common in thesaurus term lists. Having converted Sears' list to a thesaurus makes it more feasible for a user to use combinations of terms in searching, so increases its effective specificity.

A second thesaurus, shown in Appendix K, is extracts from the INSPEC thesaurus published by the Institution of Electrical Engineers (IEE) in the UK. INSPEC contains about 8500 terms with scope limited to the fields of electronic engineering and information technology, so is more specific than Sears'.

Notice first of all that INSPEC is different from Sears'. In Sears', "information systems" is a narrower term than "information science". In INSPEC, both are narrower terms of "computer applications". The narrower terms for "information systems" in INSPEC are largely information systems in particular fields, whereas in the Sears' thesaurus they are much broader in coverage. The narrower terms under "computer software" are very different between the two systems.

The INSPEC thesaurus is not a strict hierarchy in a different way from Sears'. Notice that "database management systems" has two different broader terms, "file organization" and "information systems"; while Sears' deviation noted above was a shortcut in the hierarchy. Sears' thesaurus is close to being a tree, while INSPEC is a directed acyclic graph, where one can reach a more specific term from very different broader terms. INSPEC shares this characteristic with the ACM subject list of Appendix I, as noted in the preceding section.

One way in which a thesaurus differs from a subject list is that the connections among terms are simple "see also" links in the subject list, but have the richer "broader term – narrower term – related term" structure in a thesaurus. Some systems further classify the relationships, for example using the subtype, part/whole, and set-instance relationships discussed in Chapter 8.

A few systems have a very rich taxonomy of types of relationships, for example the Medline system supporting the US National Library of Medicine's collection of medical-related journal articles. Medicine has a highly standardized set of practices and a correspondingly standardized vocabulary for its research publications. It therefore makes sense to have a large number of specialized relationships among terms so large that the specialized relationships are themselves hierarchically organized. A few examples are shown in Figure 12.1.

There are a large number of thesauri available on a more-or-less commercial basis, which can be adapted to particular purposes. For example, Oracle has a text

adverse effects
  poisoning
  toxicity
analysis
  blood
  cerebrospinal fluid
  isolation and purification
  urine
anatomy and histology
  blood supply
  cytology
    pathology
    ultrastructure
  embryology
    abnormalities
  innervation

Figure 12.1 Extract from the Medline subheading "explosion".

retrieval product called ConText, which includes a thesaurus containing one million terms, 100,000 themes, and 2000 top-level concepts, in four levels of hierarchy. The particular thesauri discussed in this chapter are also all available. Thesauri are closely related to faceted classification systems as described in Chapter 11.

Building a thesaurus

Like a large classification system, a large thesaurus is a collection of terms big enough that it requires an information system to keep track of it. A classification system has a schedule. The list of index terms is the corresponding component of a thesaurus. A thesaurus also needs its own index, for the same reason as a classification system, to support both the indexers and the users. A thesaurus does not generally include a notation, however, as it is not used for labelling or locating objects.

Design of the list of terms follows a similar procedure to the design of a classification system as described in Chapters 8 and 11. It is necessary to understand who is expected to use what information for what purpose, and to understand what resources may be economically devoted to the development project.

As described in the "Design requirements" section at the beginning of this chapter, the system must be exhaustive over its domain and specific enough to sufficiently differentiate its objects. The requirements for overview and browsing determine the parameters of the required hierarchy.

The actual terms developed and the process for obtaining them depend on the particular domain being indexed, the extent to which previously existing terms can be obtained and adapted, and the extent to which automated procedures can be used (discussed in more detail in Chapter 4 and below). Recall that even if the most specific terms are found by an automated process, the systematization into a hierarchy and the specification of relationships among terms must still be done more or less manually.

Finally, like all information systems, the thesaurus must be reviewed and tested before it is deployed, and procedures must be put in place to maintain it.

The demands on the index for a thesaurus are much greater than for a classification system. The index for a large classification system is used primarily by professional indexers. The users of the classified information will generally use the thesaurus to find the information they are seeking, rather than the classification index. Therefore, a good user interface for the thesaurus index can make a great contribution to the usefulness of an information service.

A thesaurus consists of a collection of terms associated by relationships. One way of publishing a thesaurus is simply an alphabetic list of the term–relationship complexes, as in Appendices J and K. This representation makes it difficult to see the relationships among the terms. However, since these systems are now computerized, it is possible to dynamically create alternative representations which emphasize different aspects. For example, Figure 12.2 is a representation of the INSPEC system of Appendix K showing the hierarchy.

A hierarchical representation like this can be dynamically expanded and contracted by the user. It suppresses the related-term relationships, however, and

```
Computer applications
    Information science
        document delivery
        ...
        information services
            information networks
            information resources
        information storage
        information use
        vocabulary
    information systems
        database management systems
            active databases
            ...
            visual databases
        digital libraries
        ...
        traffic information systems
```

Figure 12.2 Hierarchical representation of INSPEC thesaurus.

also the fact that some terms have multiple broader terms (the expansion of "database management systems" will appear also under "file organization" as well as under "information systems" as in the figure).

The related-term relationships can be shown in a concept map such as that described in Chapter 7. These networks of relationships can be quite large, so that visualization techniques like fish-eye views, also described in Chapter 7, can be useful. If the relationships are typed as in Medline, one can dynamically restrict the visualization to links of selected types. It is often feasible to annotate the index with the number of objects described by each descriptor. There are many possibilities.

Automated construction of thesauri

As discussed in Chapter 4, it is often possible to take a literary warrant approach and construct a list of descriptors by a statistical process on the text associated with the collection of objects. This list can be either accepted as it is created, or subjected to an editing process.

This list is not, however, a thesaurus. At best it is the list of most specific terms. In order to turn the list into a thesaurus, the relationships among the terms must be identified, and usually more general terms added since they often do not appear in the documents at all.

Sometimes it makes sense to create the more general terms by a theoretical analysis (remember that there are many fewer more general terms than most specific terms). Sometimes an existing more general thesaurus can be adopted or adapted to the purpose. It then becomes necessary to associate each most specific term with one or more of the more general terms. If a particular term does not fit well into the more general system, either the more general system must be adapted or the poorly fitting term discarded. It may be necessary in that case to discard also the document from which the poorly fitting term came if it is not otherwise indexed.

Once the automatically generated terms are attached to the more general terms, it becomes easier to identify synonyms and establish a set of preferred-term relationships, since all the synonyms should in principle be associated with the same broader term. Related-term relationships among the most specific terms are more difficult to establish, since it is necessary to compare each preferred term with every other preferred term to determine whether a relationship exists that should be recorded. It may be adequate to limit related term relationships to the broader terms in the human-constructed part of the term list.

If a literary warrant approach to building the thesaurus is desired, then it makes sense to use the computer to perform some tasks in order to reduce costs. The two key tasks are:

- *relate* – identify the terms which should be related, and what those relationships are;

- *group* – identify the terms which should be grouped under broader terms not in the original term list.

The major technical problem in both cases is that there are a large number of most specific terms, each of which must be compared with every other. It would greatly reduce the human effort needed to complete the thesaurus if programs could be written to reduce the number of comparisons which must be made.

We know that the practical impossibility of achieving 100% precision and 100% recall makes it very unlikely that we could completely automatically perform these tasks. However, we also know that some relatively simple computational procedures can significantly assist humans in these sorts of tasks. In particular, we saw in Chapter 3 that a useful approximation to the intuition that some documents are nearer to each other than others was a measure of nearness based on the proportion of shared terms, which is the basis for the query languages used in most Web search engines.

We also observed in Chapter 3 that some query languages allow the user to specify not only that two words occur in the same document, but within a specified number of words apart (proximity). The intuition here is that words which are related semantically often occur together in the same sentence or same paragraph. Since what we are looking for now is measures of similarity among words, it makes sense to look for pairs of words that occur close together in documents in the collection.

There are a large number of particular measures of term nearness based on this general idea, which are beyond the scope of this text to canvas. However, in general the measure gets larger (nearer) the more often the two words co-occur, the nearer they are to each other when they co-occur, and the fewer times they occur individually but not together. If two words co-occur more times in the same paragraph they are nearer. If they occur, on average, three words apart they are nearer than if they occur, on average, five words apart. If one never occurs without the other, then they are nearer than if the two often occur but rarely together. These measures generally discount words that rarely occur, since they can co-occur by accident. If a word occurs only once in the collection, then it might occur with other words purely by accident, so the estimates for rarely occurring terms are not statistically reliable.

Figure 12.3 shows a visualization of the results of this sort of calculation. The collection of documents is a set of 160 newspaper clippings covering topics relevant to information science. The nearness measure is computed by a method based on the strategy in the previous paragraph, whose details do not matter. Distance between terms in the figure is based on a sort of "gravity", which is computed from an iteration of direct and indirect relationships among the terms.

Notice near the bottom of the middle of the figure the term "portal". The nearest terms around it are "Excite", "Yahoo", "market", "Amazon", "services" "Internet" and "bandwidth". All of these terms, except perhaps "bandwidth", are semantically closely related to "portal". Excite, Yahoo and Amazon are instances of the class *portal*. Portals are used for markets and Internet services. Similarly, at the centre left of the figure there is a cluster of terms around "wap", including "xml", "markup", "mobile-phone", "year", "Telstra", "Microsoft" and "PC". WAP is a method of connecting mobile phones to the Internet, and includes a markup language based on XML. Telstra is the major telecommunications carrier in Australia. The terms "Microsoft", "PC" and "year" are less semantically related.

Also shown in the figure are a group of rays centred on the term "vertical". These rays represent the direct relationships among the terms, as distinguished from the combination of direct and indirect relationships rendered by distance. The strength of direct relationship is shown by a colour (not reproduced). There is a particularly strong direct relationship between "vertical" and "portal", derived from a number of articles which discuss vertical portals. The term "vertical" is not near "portal" because the latter term occurs much more often, and has in aggregate much

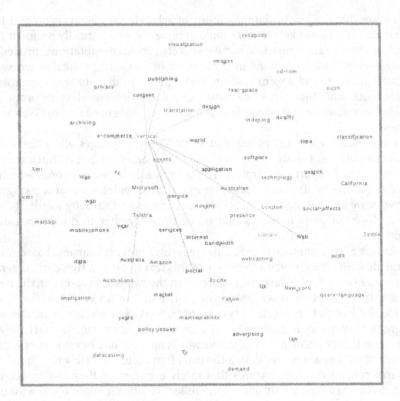

Figure 12.3 Relationships among terms.
Source: Andrew Smith, Leximancer, August 2001.

stronger relationships with terms like "Excite" and "Yahoo". The direct
relationships help to relate the less frequently appearing terms with the more
frequently occurring, giving a second method of identifying relationships.

In both cases, the statistical process greatly reduces the number of pairs of terms
which must be examined to identify related terms, thereby helping with the task *Relate*.

Of course, not all terms near each other have close semantic relationships, but
most terms with close semantic relationships are close in one or the other measure.
This is a reflection of the general problems of precision and recall introduced in
Chapter 3. The strategy illustrated in the figure seems to have pretty good recall, at
the expense of some precision.

The nearness might also help with task *Group*, identifying the terms which
should be grouped under broader terms. For example, the terms "markup" and
"xml" occur together at the far centre left. The latter is an instance of the former.
Not far away are "Telstra", and "mobile phone" which suggest a broader term, say
"communications".

Some of the other groupings of terms are less convincing. This experiment was
run on a small number of documents compared with the number of descriptors
being analyzed, so is not very statistically reliable. A larger collection would be
expected to give more consistent results.

These sorts of tools are still experimental, but many people are working on these
problems.

Descriptors as primary organizers of a collection

Historically, collections of descriptors have been used to access material in collections, but the collections have generally been organized by classification systems. The classifiers define the set of places that the objects can be. A book in a library is on the shelf indicated by its call number, even though the catalogue allows the user to search by subject descriptor terms.

However, the same is not necessarily true in fully-computerized systems such as search engines or online product catalogues. In these systems, each object is allocated a place by a process hidden from the user – say the next free disk block. The system records the address in a table with the object's public identifier. When someone finds a document from the descriptor index, the system does a further lookup in the storage directory, finds the address and retrieves the object. When an object is deleted, the storage it occupied is added back to the free list.

The advent of automated warehouses has made such systems possible even where the objects are physical. There are libraries with automated stacks, where a book is placed in the next free slot and its location recorded along with its identifier. The book is retrieved by an automated process and delivered to the reader. The location of the book is then updated as being "in circulation". When the book is returned, the automatic machinery simply assigns it to a probably different slot, and the new address is recorded.

Further, recall from Chapter 9 that sets of descriptors can define a space using a variant of hamming distance. Here, the location of the object in information space is specified by the complete set of descriptors attached to it. A query also defines a region in this space, and documents responding to the query are those whose locations are within the region defined by the query.

There is much that can be done with this view of things, but to further explore this experimental field would take us too far from current practice for the present text.

Ontologies

Closely related to the thesaurus is another class of information object called ontology. An *ontology* is a collection of terms designed to specify and structure the domain of an information system of some kind. The field is relatively new, and comes from diverse sources. A conceptual model of an information system expressed as an entity–relationship–attribute diagram or UML class diagram includes an ontology, since it specifies what sorts of things the information system is about. So also is the basic set of terms and operations specifying what can be done by a robot moving about in two dimensions.

Some ontologies are simply collections of terms, often class names. An example of this type is the Bib-1 set of attribute names appropriate to bibliographic collections, which is associated with the Z39.50 information retrieval protocol (see Figure 12.4).

Bib-1 is a collection of names of attributes, therefore a collection of class names. Like any database, any particular collection searchable using Bib-1 attributes has in addition a large number of specific terms which are instances of the various classes. These instance terms are not part of the ontology.

Personal name	Dewey classification	MESH subject
Corporate name	UDC classification	PA subject
Conference name	Bliss classification	LC subject heading
Title	LC call number	RVM subject heading
Title series	NLM call number	Local subject index
Title uniform	NAL call number	Date
ISBN	MOS call number	Date of publication
ISSN	Local classification	Date of acquisition
LC card number	Subject heading	Title-key
BNB card number	Subject Rameau	
BGF (sic) number	BDI index subject	
Local number	INSPEC subject	

Figure 12.4 Use attributes from Z39.50-1995 Appendix 3.

Notice that Bib-1 contains several sets of similar terms. There are four different card numbers, four classifications, four call numbers and three dates in the extract presented in the figure. Even though we, as human readers, can see the similarity among the terms and infer relationships among them, it is important to keep in mind that to the information system all the terms are text strings, so that "LC call number" is as different from "NLM call number" as is "Title-key".

One way to achieve a richer ontology is to classify the terms using, for instance, the classes just mentioned. For example, we could use the deeply hierarchical ACM term list in Appendix I as an ontology.

Some ontologies take explicit note of the structure of these kinds of relationships, by specifying procedures for constructing terms rather than the terms themselves, in a way similar to the Document Type Declaration of XML discussed in Chapter 7. Expressed in a related formalism called *Backus–Naur Form* (BNF), we might have definitions such as those in Figure 12.5. The term on the left-hand side of the "::=" is replaced by the complex on the right. Terms not enclosed in angle brackets are terminals, as in the DTD syntax.

With this sort of term generator type of ontology, one can more easily write programs that treat all classifiers in a similar way, all dates similarly, and so on.

Finally, we will look at WordNet, which is a large collection of English-language words in linguistic classifications (nouns, verbs, adjectives, and adverbs) with a rich collection of broader-term/narrower-term relationships. Words generally have synonym and antonym relationships. Synonyms are taken as equivalent, so the basic unit in WordNet is a set of sets of synonyms, called *synsets*. A search in WordNet for the word "search" results in Figure 12.6.

```
<qualified-name>::= <name qualifier> name
<name qualifier> ::= personal | corporate | conference
<qualified classification> ::= <classification qualifier> classification
<classification qualifier>::= Dewey | UDC | Bliss | Local
```

Figure 12.5 Fragment of a term generator for Bib-1.

The noun "search" has 5 senses in WordNet:

1. search, searching, hunt, hunting – (the activity of looking thoroughly in order to find something or someone)

2. search – (an investigation seeking answers; "a thorough search of the ledgers revealed nothing" or "the outcome justified the search")

3. search, lookup – (an operation that determines whether one or more of a set of items has a specified property; "they wrote a program to do a table lookup")

4. search – (the examination of alternative hypotheses; "his search for a move that would avoid checkmate was unsuccessful")

5. search – (boarding and inspecting a ship on the high seas; "right of search")

The verb "search" has 4 senses in WordNet.

1. search, seek, look for – (try to locate or discover, or try to establish the existence of; "The police are searching for clues"; "They are searching for the missing man in the entire county")

2. search, look – (search or seek; "We looked all day and finally found the child in the forest"; "Look elsewhere for the perfect gift!")

3. research, search, explore – (inquire into)

4. search – (subject to a search; "The police searched the suspect"; "We searched the whole house for the missing keys")

Figure 12.6 WordNet entry for "search".

There are nine entries for "search", each of which has a different meaning, called a *sense*. Five of these senses are nouns and four verbs.

There are several types of broader term/ narrower term relationship, as well. One expresses a subtype of relationship, as shown for "search" in Figure 12.7. These relationships have special names *hypernym* and *hyponym*. The term "investigation" is a hypernym of "search", while "search" is a hyponym of "investigation".

Verbs can have another type of broader term/narrower term relationship called *troponym* (a troponym of a verb is a way of doing that action), as in Figure 12.8.

A third type of broader term/narrower term relationship in WordNet is the part-whole relationship. An example, for one of the senses of "library", is shown in Figure 12.9. Again, there are special names for these relationships. The term "reading room" is a *meronym* of the term "library", while "library" is a *holonym* for "reading room".

WordNet is not strictly an ontology in the sense of the other examples used, since it was intended as a tool for research in language understanding, rather than a set of class names for information systems operation, but it is often used in combination with other ontologies, and also provides an extensive example of a system of terms with a rich set of relationships.

search – (an investigation seeking answers; "a thorough search of the ledgers revealed nothing" or "the outcome justified the search")
> => investigation, investigating – (the work of inquiring into something thoroughly and systematically)
>> => work – (activity directed toward making or doing something; "she checked several points needing further work")
>>> => activity – (any specific activity or pursuit; "they avoided all recreational activity") act, human action, human activity – (something that people do or cause to happen)

Figure 12.7 Hypernyms of "search" in WordNet.

search, seek, look for – (try to locate or discover, or try to establish the existence of; "The police are searching for clues"; "They are searching for the missing man in the entire county")
=> dredge, drag – (drag, usually the bottom of a body of water)
=> finger – (search for on the computer)
=> grope, fumble – (feel about uncertainly or blindly; "She groped for her glasses in the darkness of the bedroom")
=> divine – (search by divining, as if with a rod, of underground water or metals)
=> browse – (look around casually and randomly, as through files and directories on a computer)
=> want – (hunt or look for; want for a particular reason; "Your former neighbour is wanted by the FBI"; "Uncle Sam wants you")
=> scour – (examine minutely; "The police scoured the country for the fugitive")
=> seek out – (look for a specific person or thing)
=> quest for, go after, quest after, pursue – (go in search of or hunt for; "pursue a hobby")
=> maraud – (search for plunder)
=> fish, angle – (seek indirectly; "fish for compliments")
=> ferret grub – (search busily; "ferret out the truth")
=> feel – (grope or feel in search of something; "He felt for his wallet")
=> grope for, scrabble – (feel searchingly; "She groped for his keys in the dark")
=> shop, browse – (shop around; not necessarily buying)
=> comparison-shop – (compare prices for a given item)
=> antique – (shop for antiques; "We went antiquing on Saturday")
=> window-shop – (examine the shop windows; shop with the eyes only)

Figure 12.8 WordNet 1.6 results for "Troponyms (particular ways to this)" of verb "search" Sense 1.

> library, depository library – (a facility built to contain books and other materials for reading and study)
> HAS PART: carrel, carrell, cubicle, stall – (small individual study area in a library)
> HAS PART: reading room – (a room set aside for reading)
> HAS PART: stacks, set of bookshelves – (an extensive arrangement of bookshelves in a library where most of the books are stored)

Figure 12.9 WordNet 1.6 results for "Troponyms (particular ways to this)" of verb "search" Sense 1.

Key concepts

Non-classification indexing systems include **subject term** systems, where each subject term describes a complete concept; and the **thesaurus,** where several terms may be needed to describe a concept. A thesaurus is a list of terms with relationships among them, especially **broader term/narrower term** and **related term.** An **ontology** is similar to a thesaurus, but generally used as a basis for information systems rather than for document retrieval.

Further reading

Further information on subject lists and thesauri can be found in Rowley (1992: Ch. 16). A number of database manufacturers have text database products including a thesaurus. More information can be found on the manufacturer's Web site. In particular Oracle has a product called ConText. More information on WordNet is in Fellbaum (1998).

Formative exercise

Find a published set of descriptors in an area with which you are familiar.

- Is it best described as a subject list, thesaurus or ontology?
- What is its scope?
- Is it specific enough for usable queries?
- How hierarchical is it?
- Does it have an index?
- Does it support cross-reference?
- If it is hierarchical, are its hierarchical relationships typed?
- Was it built using automatic methods?
- How is it maintained?

Tutorial exercises

1. Develop both a subject index and a thesaurus to support a collection of documents including the following. Compare the two.

No.	Title	Author
0001	Aboriginal health and history power and prej	Hunter Ernest
0002	Advanced emission controls for power plants	
0003	Advanced coal-based power generation technic	Bharucha Noshir
0013	The application of expert systems in the powe	
0014	Asia-Pacific Economic Co-operation Working Gr	
0018	Coal solid foundation for the world"s electr	
0023	Constitutional law in the Middle East the em	Mallat Chibli
0024	Desulphurisation 3	
0030	Electric power technologies environmental ch	
0038	Energy taxes and greenhouse gas emissions in	McDougall R.A.
0039	Energy policies of Romania survey	
0040	Environmental aspects of nuclear power paper	
0044	Flooding Job"s garden	
0047	Gas engines for co-generation papers present	
0050	High-power GaAs FET amplifiers	
0065	Iron, gender, and power rituals of transform	Herbert Eugenia W
0072	Judicial power and the charter Canada and th	Manfredi C
0074	Last rights death control and the elderly in	Logue Barbara J.
0075	Law, liberty, and justice the legal foundati	Allan T. R. S.
0076	The legislative process in the European Commu	Raworth Philip
0078	Maritime change issues for Asia	
0081	Museum of contemporary art vision & context	Murphy Bernice
0084	Nuclear power plant safety standards towards	
0085	Off-site nuclear emergency exercises proceed	
0087	A peaceful ocean? maritime security in the P	
0099	The power of one	
0100	Power electronics circuits, devices, and app	Rashid Muhammad
0101	Power electronics semiconductor switches	Ramshaw R. S.
0102	Power MOSFET design	Taylor B. E.
0113	Purchasing power parities and real expenditur	
0119	Report on a review of police powers in Queens	Queensland
0120	Report on review of independence of the Attor	Queensland
0121	Report on review of independence of the Attor	Queensland
0125	Rural and remote area power supplies for Aust	
0126	A short course on design of transmission line	Short Course on Desi
0130	Spectrum estimation and system identification	Pillai S. Unnikrish
0131	Spectacular politics theatrical power and ma	Backscheider Paula
0132	Up against Foucault explorations of some ten	
0135	Visitor centres at nuclear facility sites pr	
0137	Work management to reduce occupational doses	
0138	World energy outlook to the year 2010	

Open discussion question

Compare and contrast the use of a classification system versus use of a subject index system for structuring an information space. Do the two do a similar job? How are they different? Does it make sense to use both? If so, how do they interact from the user's point of view? Consider particular examples.

13 Visualization

We look at how classification systems can be used as the basis for visualization of an information space.

Information as a space

A space is a collection of places where things can be. In Chapter 5 we observed that humans operate in a wide variety of spaces, from geographical space through urban space to a building seen as a space. Our ability to make use of these spaces is enhanced by a wide variety of information structures such as maps, directories, addresses and plans.

One of the things that can be in a space is an observer. Sometimes the observer is immersed in the space, as we are in real three-dimensional space, and sometimes the observer sees the space from the outside as we do when looking at a two-dimensional space. If we are immersed in a space we can't see all of it at once, almost by definition. If we are outside the space, sometimes we can see all of it and sometimes only a part. Maps are two-dimensional objects which enable us to stand outside the space they represent so we can see features and patterns we are interested in.

We often use the metaphor of a window to metaphorically frame what we can see in either case.

Not only are things in space, but things can move in space. In particular, the observer can move. Movement of the observer is moving the metaphorical window through which we see the space. If we are outside the space and can't visualize it all, then we can think of ourselves as inhabiting a space with an extra dimension, so that we can be metaphorically above the lower-dimensional space we are observing. If we move away from the observed space, we see more of it but at a lower resolution; while if we move towards the observed space, we see at a higher resolution but less of it. This movement is often called zooming out and in. We can also move horizontally to the space, so we can see the same amount of it at the same resolution, but different regions. Scrolling in a hypertext window is this sort of movement.

In Chapter 5, we used the metaphor of space to help us organize a hypertext, a large structured document. Here, we will make use of the concepts of semantic dimensions introduced in Chapter 9 to examine a few of the ways we can use the space metaphor to understand large collections of individual objects.

Navigation

In our discussion of information spaces, we are always assuming that the user wants to find some information. They may or may not know what they are looking for at the start of their search, but if they finish satisfied, they have found out something. Following the navigation metaphor, we call the information they find out *destination information*.

However, there is generally much more information in a visualization of an information space than destination information. In real space, if I walk out of my hotel and want a newspaper, I can see down the street to the left a shopping district. I think I may be able to find a paper in that direction, so I head for the shops. Once in the district, I can see among all the signs on the shops a sign that says "Newsagency". I can probably find my paper in there. I can now see the front of the shop and can make out a doorway, so I go in. To the left of the cashier, I see several piles of newspapers, so I go there. The top paper in each pile has the paper's name on it, so I look for a pile whose top paper says "The Australian". Finding that pile, I pick up one of the papers, and now have what I want.

But in navigating to what I wanted, I made use of a large amount of information – the recognition of the shopping district, the "Newsagency" sign, the recognition of the doorway between the shop windows, the piles of papers, and the label "The Australian". We call all of this information *navigation information*. A good visualization of an information space will provide sufficient navigation information to help us navigate to the destination information we are seeking.

Computerized information spaces vary considerably in their provision of navigation information. For example, suppose I am looking for an e-mail list for the students in one of my classes. The destination information might look like Figure 13.1: a list of students' names together with their e-mail addresses.

Some systems, for example a bare UNIX platform, might present the user a screen like Figure 13.2 – a bare prompt. There is no navigation information at all. The user must know the right command, say "cat is-mailing", to obtain the destination information of Figure 13.1.

Recognizing this deficiency of command-line interfaces, most systems have for many years had a menu interface something like Figure 13.3, the menu from the UNIX e-mail handler *pine*. This menu has considerable navigation information. There is a list of options, which is a nominal system in the language of Chapter 9, together with instructions for making a choice. I can see that the path to what I want is very likely to be via the choice "A Address Book".

Besides the navigation information, there is also some destination information on this screen. Both at the top and bottom it says that there are 74 messages in the "In box", which tells me that if I wanted to check my e-mail, there are messages and how many. This is a representation of destination information by aggregation, in this case by *count*.

On the other hand, note that in Figure 13.1, there is destination information, but no navigation information. I have what I told the system I wanted, but have no assistance in finding anything more. Most contemporary Web-based interfaces have a better balance between destination and navigation information. Consider Figure 13.4, a screen from the university's library catalogue after I entered the name "Salton" in the "author" field of a search form. I am looking for a particular book. The screen does not give me all of what I want, but some – it indicates by showing in entry number 2 that there are seven books whose author is Gerald Salton, so that

Surname	Given names	e-mail address
Aberdeen	Jacqueline	JABERDEEN@library.uq.edu.au
Akamatsu	Hitomi	hitomi0829@yahoo.com
Bosschieter	Joshua James	s350747@student.uq.edu.au
Brown	Stephen Marcus	s358837@student.uq.edu.au
Brown	Stuart Douglas	s333553@student.uq.edu.au
Chen	Wei-Chi Wiki	wikiocean@hotmail.com
Cheng	Michael Li-Chyun	s346155@student.uq.edu.au
Chia	Po San Fione Anne	s806791@student.uq.edu.au
Coleman	Laurence Sean	s366558@student.uq.edu.au
Cross	Jaymie Eliza	jcross80@email.com
Cross	Matthew	mrc@bigpond.net.au
Cruz	Maria Ramona V.	s360506@student.uq.edu.au
Cueva	Francis Michael	i_c_abigfanni@hotmail.com
Foong	Chee Hong	chee_hong@rocketmail.com
Gan	Shen-Lene	lenelene@hotmail.com
Goh Chuen	Ping Alvin	s802261@student.uq.edu.au
Gooding	Glenn Michael	s355764@student.uq.edu.au
Havazvidi	Liberty Mushangwe	liberty28@hotmail.com
Ho	Jonathan Kia Hwai	s341419@student.uq.edu.au
Ho	Wai Kit James	s803922@student.uq.edu.au
Holmes	Matthew Ryan	s369440@student.uq.edu.au
How	Envin Adrian	tak_s@hotmail.com
Jadin	Raphael Dominique	s343371@student.uq.edu.au
Jardine	Micah John	s349784@student.uq.edu.au
Kang	Eu Gene Eugene	wired@genes.pl.my
Kee	Keam Hong	khkee@yahoo.com

Figure 13.1 Sample destination information – a list of e-mail addresses.

the information I want very likely exists, and suggests I proceed by clicking on the author's name.

When I do that, I get the screen in Figure 13.5. This contains the details of the book I was looking for, plus a large amount of navigation information. I can pass to

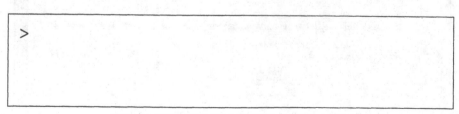

Figure 13.2 A bare UNIX screen.

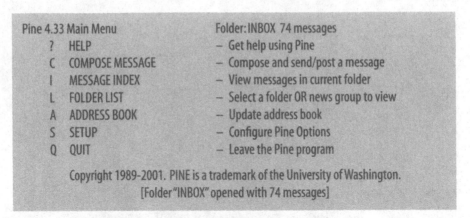

Pine 4.33 Main Menu   Folder: INBOX 74 messages
        ?   HELP   – Get help using Pine
        C   COMPOSE MESSAGE   – Compose and send/post a message
        I   MESSAGE INDEX   – View messages in current folder
        L   FOLDER LIST   – Select a folder OR news group to view
        A   ADDRESS BOOK   – Update address book
        S   SETUP   – Configure Pine Options
        Q   QUIT   – Leave the Pine program

        Copyright 1989-2001. PINE is a trademark of the University of Washington.
        [Folder "INBOX" opened with 74 messages]

Figure 13.3 The *pine* menu.

other books by the same author, to other books with the same subject descriptors, or to the call number index. I can also pass to the next or previous entry on the screen of Figure 13.4.

A good visualization system for an information space will maintain an appropriate balance between navigation and destination information, and will never leave the user stranded by providing no navigation information at all.

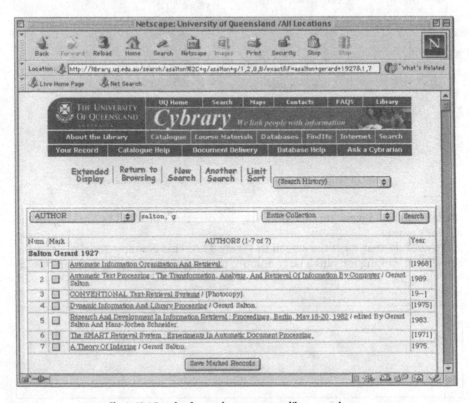

Figure 13.4 Result of an author query on a library catalogue.

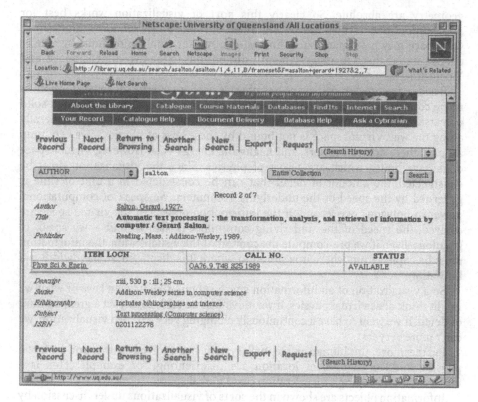

Figure 13.5 The library catalogue entry for a particular book.

We now turn to a few specific examples of visualization systems and give some design guidelines for these important special cases.

Keeping close to real space

Humans have evolved in real, three-dimensional space so our information-gathering and navigational perceptual mechanisms are designed to work best in that environment. One way to design an information visualization system, therefore, is to keep as close as possible to real space. Virtual reality is an extreme example of this.

Real space has a geometry, that described by Euclid in ancient times, in either two or three dimensions. A two-dimensional space lets us get an overview, while a three-dimensional space doesn't, as noted above, so we will confine ourselves to two dimensions and use the third for zooming in and out. Humans can make fine distinctions of locations in space, so our two-dimensional visualization can have many places. The Euclidian geometry describes our ability to move and see density in space.

Therefore, a visualization close to real space works best if we can choose semantic dimensions which are fine-grained and linear. Because we see real space as unitary, the visualizations work best if the semantic dimensions used for the geometry are absolute. So clearly this sort of visualization works best for information spaces which have fine-grained absolute linear semantic dimensions. Our space may have many such dimensions, so the visualization must be able to choose among them, to assign semantic dimensions to the two-dimensional X and Y coordinates.

Besides a geometry, real space has physics, which governs movement. We must move continuously – there are no hyperspace ports. We can't move through solid objects. Movement costs energy, at least in starting and stopping and generally along the way. Virtual space is not bound by these laws of physics – we can jump around, there are no barriers, and negligible energy cost. However, virtual space does have limits, which are a sort of physics. Since the visualization must be constructed, the amount of detail that can be constructed in a unit of time is governed by the speed of the underlying computers. We speak of computational power in polygons per second. If the information objects of the virtual space are derived, the speed of the underlying computers is also a limit – we speak of gigaflops. Even if we pre-compute the content of the visualization, the information must be copied from where it is stored to where it is displayed, so bandwidth is a limitation.

So a visualization of an information space is bound by its own laws of what we might think of as virtual physics. If we move fast we can't construct a great amount of detail. If we want to have a continuously changing very detailed visualization, we can't more very fast.

Furthermore, the human perceptual system has evolved in the world of real physics. Sudden changes of location are disorienting, for example. This is a motivation for keeping a visualization close to real space.

Information objects are shown in the sorts of visualizations under discussion by points in space. The fact that we can have many semantic dimensions in our information space but can only represent two of them as spatial dimensions is a severe limitation – to understand the information space we often need to be able to visualize many dimensions at once. Consider, for example, Figure 13.6. There, the information is visualized as points in a two-dimensional space, but the points are differentiated by the method of Chernoff faces, incorporating more information in the visualization. (It is not indicated what data is being shown, as the figure has no legend – the point is the method of visualization, not especially what is being visualized.)

We can see from the figure that there is a weak sort of association between the semantic dimensions represented in the X and Y coordinates, since the upper left and lower right of the window are empty. The points themselves tell us more. The lower left points are all represented by smiling faces with exaggerated jaws and small eyes and noses. As we move up and to the right, the faces become squarer, with staring eyes and long noses. Further along, the faces become rounder, with more smiles. The face labelled *a stranger* is very different from its neighbours, so breaks the general pattern. This point might be worth investigating – it might be erroneous data or it might be a very significant observation.

What this visualization is doing is making use of the human's ability to differentiate faces to represent more semantic dimensions – there are about eight represented by the crown and jaw outlines, length and smile/frown of the mouth, length of the nose, and size, spacing and height of the eyes. Use of the ability to differentiate objects is very similar to the use of shared words to represent document nearness. As we saw in Chapter 9, this representation of nearness appears

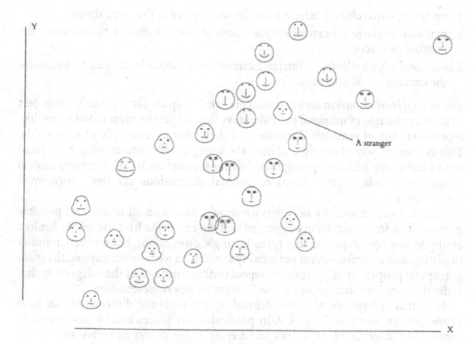

Figure 13.6 A visualization using Chernoff faces.
Source: Chernoff (1973).

in space as a geometry based on hamming distance or its relatives. For example, in the figure there are no faces with staring eyes and short noses or wide smiles. This lack represents empty space just as do the lack of points in the upper left and lower right.

Chernoff faces are not widely used, but it is very common to represent data points as objects with what we might call personalities – geometric shape, colour, shading, etc., which have the same sorts of properties. Humans might be very good at recognizing subtle differences in spatial location, but are not good at recognizing subtle differences in the personality of objects. Each feature on the Chernoff faces has a small number of distinct values. Only a small number of different shapes, sizes, colours or shadings are easily distinguished. For this reason, semantic dimensions represented as aspects of personality are generally coarse grained. Some personality types (size, density of shading) have a natural sequence and can be used for linear or ordinal dimensions. Other types (geometric shape or colour) have no natural sequence so are particularly suited to nominal dimensions.

In our virtual reality visualizations, we have seen that we can represent semantic dimensions in two general ways, either as spatial coordinates or as the personalities of the representations of the information objects. We will synthesize these lessons by defining two types of representations of semantic dimensions:

1. *extrinsic*, or space-like – the X or Y coordinate of the display;
2. *intrinsic*, or face-like – aspects of the personality of the points representing the information objects.

And three design principles:

1. *exclusion* – two objects cannot be in the same place at the same time;
2. *maximal exclusion* – extrinsic dimensions should be chosen to maximize the number of places;
3. *maximal object identity* – intrinsic dimensions should be chosen to minimize the occurrence of unique objects.

The principle of exclusion says that each place in the space can have only one object in it. The principle of maximal exclusion says that we get the most out of space-like representations of semantic dimensions if they have as many places as possible. This is another way of saying that they are fine-grained. Further, we get the most out of our space-educated perceptual abilities if nearness in the representation is semantically near, so that linear or ordinal dimensions are best represented extrinsically.

Maximal object identity says that we should have a small number of possible personalities for information objects, so that we can make best use of our limited ability to distinguish personality types in single dimensions, but our better ability to distinguish several different personality aspects. As we have seen above, the main perceptual properties of personality aspect is difference rather than degree, so that ordinal or nominal dimensions are well suited to this representation.

Note that the system of places defined by the semantic dimensions can have properties, as noted in Chapter 9. In particular, the places can be designated as *explored* or *unexplored*, as we saw in Chapter 5. But places can have all sorts of properties – we might think of them as *zones* by analogy to city planning, where the presence of an object has some particular semantic value for the user for whom the visualization is created.

Example of close-to-real-space visualization

An example will make the visualization strategies clearer. In 1993, the fire brigade in the state of New South Wales, Australia, instituted a system for recording sites in the states which stored chemicals. This system, called SCID, records what chemicals are stored, where they are stored and under what conditions, so that if there is a fire or other emergency the brigade can use the appropriate methods, take necessary precautions, organize evacuations and so on. The information in the SCID system is a natural candidate for close-to-real-space visualization.

Let us imagine that we have a general-purpose visualization system. The one we will sketch is somewhat crude, but illustrates the principles. The system is capable of showing two extrinsic dimensions and two intrinsic dimensions. The intrinsic dimensions are represented by two personality types, the geometric shape of the point and by the amount of shading of the geometric object. Each representation has three possible values, as shown in Figure 13.7.

The shading personalities have a natural order, while the shape personalities do not.

There is a variety of information associated with each chemical storage site in the SCID system. The location of the site is stored in map coordinates. The contents of the site are represented using many semantic dimensions, including type of chemical, how volatile it is, how toxic it is, how long it persists in the environment, how close it is to population and how securely it is contained. Type of chemical is a nominal dimension, while the others can be represented as fine-grained linear

Shape personalities   Shading personalities

Figure 13.7 Intrinsic dimension personality types.

dimensions. For example, persistence can be represented as hours until it is no longer toxic or security of confinement as bursting pressure of the container. However, we have seen that any fine-grained dimension can be turned into a coarse-grained dimension by grouping or thresholding, so that toxicity or security of containment can be represented as *low*, *medium* or *high*, a coarse-grained ordinal dimension.

Our visualization system allows us to simultaneously represent two extrinsic and two intrinsic dimensions, so the first problem is to decide what four dimensions we want to see, then the next is how to represent each dimension chosen. We could use the extrinsic dimensions to represent the map coordinates, but we want to be a little more adventurous, so we will choose dimensions that are not naturally spatial – say toxicity, flammability, security of confinement and proximity to people. We need to represent two of these extrinsically and two intrinsically.

Of the four dimensions chosen, two (toxicity and flammability) can be thought of as properties of the chemicals stored and the other two (proximity and security) can be thought of as representing the risk of serious damage. What the fire brigade would like is that sites storing highly toxic and highly flammable chemicals were either securely contained or far from population, preferably both. It is plausible that they would be concerned with risk in a coarse-grained way, since the measures they can take are discrete (evacuate or not, say), while the measures taken to deal with toxicity or flammability might be more flexible. Therefore we will choose to represent toxicity and flammability as extrinsic dimensions, but security and proximity as intrinsic dimensions. Both of these last are ordinal, so we will assign arbitrarily security to the geometric shape personality and proximity to the shading personality.

Visualization of this selection of dimensions according to the representation choices above might look as Figure 13.8.

At the bottom of the picture is the legend. The legend shows that we are considering six semantic dimensions, of which we have omitted two (persistence and volatility) from the visualization. At the left of the legend we show that toxicity is represented as the vertical extrinsic dimension, while flammability is represented as the horizontal. At the right we show the personalities used to represent security and proximity. A site with low security of confinement and low proximity to people is represented as a white oval, while a site with high security and high proximity is represented as a black rhombus.

The main display is framed by two navigation bars. The horizontal bar has at the left "2" and at the right "9" (*range indicators*) signifying that the lower extreme is 2 units of flammability while the upper extreme is 9 units. Similarly, the vertical navigation bar range indicators show that the lower extreme of toxicity represented is 4 units while the upper extreme is 7 units. Not all the sites have values of the extrinsic attributes within that range, so not all sites are shown in the visualization. The display shows that there are 6 sites with less than 2 units of flammability and 4 sites with more than 9, by the numbers "6" and "4" to the left and right of the lower

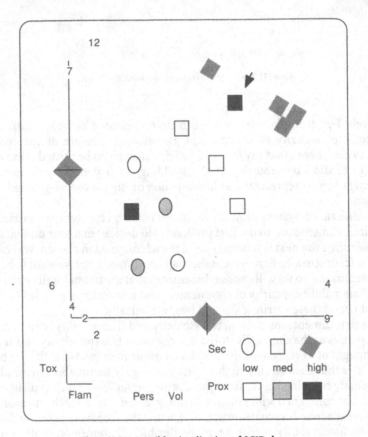

Figure 13.8 A possible visualization of SCID data.

part of the display. Similarly, the display shows that there are 2 sites with less than 4 units of toxicity and 12 sites with more than 17. The triangle pairs at the midpoint of each navigation bar are click-sensitive and will move the window in the direction indicated. If the window moves, the sites visualized change, the range indicators change and the counts of sites not shown are updated.

Both navigation and destination information are used in the visualization. The navigation bars, range indicators, the movement triangles and the legend are all navigation data. The sites visualized are destination data, as are the counts of sites not shown.

Turning to the sites visualized, we can see that toxicity and flammability are correlated to a degree, in that there are no highly toxic, low flammable sites, nor any high flammability, low toxicity sites. The five sites at the lower left have either low or medium security of confinement, but are relatively low toxicity and low flammability. As we get more toxic and more flammable, security of confinement increases and proximity to people decreases. Sites at the most toxic and most flammable end are relatively close to people, but are generally highly securely contained. The exception is the site pointed out by the arrow (not part of the display), which is very close to people but has only medium security. This "stranger object" is an exception to the pattern, and probably warrants investigation. The visualization system would allow the user to click on that object to bring up its details.

Note that the extrinsic dimensions define a space larger than the window, so that some of the destination information is the number of unseen objects in the different possible directions of movement. Another way to express this is to say that the window defines a selection of the objects on the attributes chosen as extrinsic dimensions in the visualization. In this example, however, that selection exhausts the population in the window. The intrinsic dimensions include all possible values of the attributes chosen, so that there is no further selection.

It would be possible to visualize a selection using the intrinsic dimensions by, for example, drilling down. There are only three possible values for each of the two intrinsic personalities, so that the attributes were represented at a very coarse granularity – *low, medium, and high*. We could further subdivide say the *high* range into three parts and not represent the objects having attribute values in the *low* or *medium* ranges. If the visualization platform provided this facility, it would be necessary to implement mechanisms for moving in intrinsic space as well as for moving in extrinsic space.

Finally, the space itself can be zoned. We might shade the upper right quadrant as a region where the combination of toxicity and flammability warrant either secure containment or distance from population.

Discussion of close-to-real-space visualization design guidelines

Our three design guidelines for close-to-real-space visualization are derived from real space physics and human physiology, but have no physical necessity in virtual space – this is why they are called design guidelines rather than laws.

In particular, the principle of exclusion arises from the fact that places in real space are infinitesimally small. In information spaces, places are often able to hold many objects. One way to visualize spaces where the places are larger than objects is to violate the principle of exclusion, for example in the cholera map of Figure 10.6. The information objects include deaths, which are located at addresses. Many people live at any given address, so there are many cases of multiple deaths at a single address, particularly along Broad Street. These are represented by multiple dots stacked one on top of the other. This tactic works well in that it supports the main rhetorical purpose of the map which was to show the distribution of density of deaths in relation to the location of wells.

Another way to visualize information spaces with large places is to represent the collection of objects in each place by an aggregation, say, number of objects or the average of some linear dimension. These aggregations become virtual objects which can be visualized as we have seen. Indeed, we can interpret the cholera map as keeping the principle of exclusion if we interpret the stacks of dots as a way of representing a linear semantic aggregate, in much the same way as some visualizations use circles of different sizes or intensity of shading as intrinsic dimensions to represent linear semantic dimensions.

Visualizations supporting animation can violate the principle of maximal object identity, especially if the information objects are successive states of some process. We can use time as another extrinsic dimension, so that small changes in semantic dimensions represented as intrinsic dimensions can become visible as gradual changes in the animation. This would also work for more abstract semantic dimensions, where a fine-grained linear semantic dimension can be represented extrinsically as time, especially if the semantic dimensions represented intrinsically show gradual changes correlated with the dimension represented by time. The SCID visualization of Figure 13.8 would work in this system, if the time dimension were used to represent the sum of toxicity and flammability and the intrinsic dimensions were as they are in Figure 13.8. The anomaly would show as a sudden blip.

Finally, advances in computer and communications technology have resulted in information spaces imbedded intimately and dynamically with real space. Active barcodes make it possible to automatically locate objects in space, while active badges make it possible to locate people. Further, mobile phone or global positioning system technology also locates people in physical space. One consequence is that it is now common to use visualization where the extrinsic dimensions represent real space and time coordinates, giving animations of the changing state of yacht races or the location of ambulances on a city street map.

Besides these overview applications, there are an increasing number of applications where mobile objects or people interact with information systems located in particular places if the object or person is located near on in those places. The ability to send advertising messages from restaurants in a country town to cars on the highway near the town around lunchtime is real.

We can also interpret the push technology of Chapter 4 and Chapter 6 in this way. A persistent object can have state – that is, the information associated with it can change over time as it passes through a process of some sort. This can be a company's periodic financial reports, the progress of a software project, or a person's pattern of Web site hits. The information object associated with the persistent object therefore changes, and can be thought of as moving through the information space. An information system can monitor particular places or regions in the information space and take some action if an object appears in its region – report to an investor a company with a financial report having a particular pattern, or to the marketing department when a software project gets to the alpha test stage, or to generate an advertising message to a person when their pattern of Web hits indicates they might be receptive to it. These applications can be seen as variations on the zone concept described above.

Here of course the visualization is not bound by human perceptual physiology, and in particular can cope with spaces of many dimensions simultaneously.

Viewing push or trigger technology as information visualization is far from the only useful way to see it, but this view can give a useful perspective.

Visualizing relative dimensions – enumerative systems

Our first look at visualization was intended for systems with faceted classification systems based on single-level absolute semantic dimensions. In Chapters 7 and 11 we have seen that very many classification systems include enumerative – relative hierarchical nominal – facets, many of them very large. How can we visualize systems like these?

Places in enumerative systems tend to be fairly large, due to the principle of uncertainty and the consequent unreliability of highly specific classes – large in the sense that they typically contain many objects. Therefore, the principle of exclusion does not apply. Nor, of course, does the principle of maximal exclusion. However, we have seen above that systems with many objects in a place can often be usefully visualized using aggregates, which have a single value for each place. This gives scope for visualizations using the principle of exclusion.

Places in enumerative systems are organized hierarchically, as in the examples of Appendices A–F. We have seen in Chapter 9 that navigation in such systems includes both moving from place to place and also hiding/showing places using drill-down and roll-up operations. The examples in the appendices have many instances of differential expansion of hierarchy. The key problem in visualizing enumerative systems is the effective use of screen (or paper) real estate – showing as much as is possible of very large systems. This brings to the fore a design principle of Edward Tufte for the destination data – *maximize the data-ink ratio, within reason*. This is to say, use the minimum of real estate for lines, labels and decoration, subject to keeping the visualization understandable to its users. Another way to state this principle is – *maximize the pixels devoted to destination or navigation data*.

A good example of a visualization tool for enumerative data is WebTOC, developed by David Nation and others at the University of Maryland. A sample visualization is shown in Figure 13.9.

The particular collection being viewed is the American Variety Stage part of the American Memory collection of the Library of Congress in the United States. This is a very large collection of ephemera, mostly in the form of digitized paper documents, pictures, sound recordings, films and text. Most of the multimedia objects were not at this time captioned, and most of the text of things like magazines was in the form of images, not machine-readable text. So the major organizing principle was the enumerative classification system partly shown in the figure. The WebTOC tool allows the user to differentially expand the classification hierarchy using conventional drill down/ roll-up navigation.

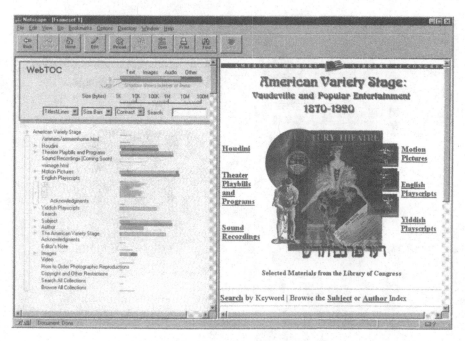

Figure 13.9 A WebTOC visualization.
Source: Nation et al. (1997: Figure 1).

Besides the main enumerative system, objects in the collection are classified using a number of small facets, one of which is the nominal system *media type*: {text, images, audio, other}, and another is the linear system given by the object's size in bytes. Since these other facets are small, the tool uses the principle of maximal object identity to visualize the contents of each place as a pair of aggregations. The upper object in each place is a bar whose length is the sum of the sizes of all the objects in the place, on a logarithmic scale – a bar twice as long as another represents the square of the number of bytes. The bar is differentiated into four coloured segments, which represent the percentage of the total size taken up by each of the *media type* foci. Below the bar is a shadow whose width shows the other aggregation, the number of items in the place. Single objects (in this case files) are visualized as lines, with length representing size and colour *media type*.

Notice that some of the files are shown without labels. Labels can be turned on and off. Turning them off gives a more compact display, showing the type, size and pattern of objects in the collection without any names. Names can be suppressed entirely, as in Figure 13.10. There, the display shows the structure, size and type of the collection without any semantics in a very dense display – a minimum of non-data ink.

WebTOC is a flexible visualization system. It can be configured for any enumerative classification system. The colour personality can be used for any small facet. For example, documents in an engineering firm could be classified as historical (related to past projects), current (related to current projects) and future (proposals and plans). A release of a complex software system could classify its modules as carried forward from previous release, modified or new in current

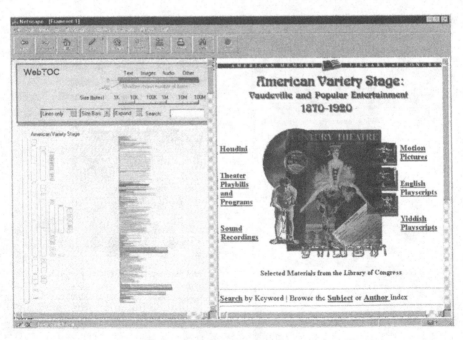

Figure 13.10 A maximally dense display from WebTOC.
Source: Nation et al. (1997: Figure 2).

release, and scheduled for change in a future release. Web sites in Yahoo! could be classified as being in one of a small number of languages.

More complex visualization systems can be built using WebTOC as a component. It is possible to represent the contents of a directory more directly than by aggregation. A simple list will do if the specificity of the places is high so there are few documents in any one. But when the place is shown rolled up to a high level of generality there may be very many objects in a class, organized in a complex subclass hierarchy. The contents of a class can be represented by sampling – a window is opened containing a random sample of the objects in the class, taken irrespective of their place in the hierarchy. Such a method is in fact used by the Library of Congress American Memory Web site.

The WebTOC display can be integrated with a selection mechanism. If the objects in the collection are described using a set of fairly general descriptors, the result of a query will often be a large set scattered through the hierarchy of the enumerative facet structuring the WebTOC display. The display can be dynamically modified to show only the classes and subclasses which are not empty.

Multiple views can be maintained and synchronized with each other. Suppose for example that the sites represented in the SCID system visualized in Figure 13.8 were classified by a facet which was the Standard Industrial Classification of Appendix A. The display of Figures 13.8 and 13.9 could be shown simultaneously. Any selection in either view could be used to immediately update the other.

Key concepts

Visualization of an information space involves **navigation data** and **destination data**, and requires a choice as to which semantic dimensions will be represented as **extrinsic (space-like) dimensions**, and which as **intrinsic (face-like) dimensions**. Various design principles apply, including the **principles of exclusion, maximal exclusion, maximal object identity**, and **minimization of non-data ink**.

Further reading

We have shown in this chapter a few methods for visualizing information spaces. The concepts of navigation and destination data, as well as the principles associated with close-to-real-space visualization come from Benedikt (1991a). The WebTOC visualization was developed by Nation et al. (1997), but there are an enormous variety of visualization techniques available. The series of books by Tufte (1983; 1990; 1997) are an excellent source. The Web site Atlas of Cyberspace (Dodge and Kitchin, 2001) shows an amazing variety of visualizations.

Formative exercise

Find several complex visualisations in an area with which you are familiar.

- If a visualization is dynamic, what in it is destination data and what is navigation data?

- Is the visualization close-to-real-space? If so, what dimensions are represented extrinsically and what intrinsically? What personality aspects are used for the intrinsic representations? Does the choice of dimensions accord with the design principles of this chapter?
- Are there classification systems and visualizations different from those described in this chapter? Do they work? Are they potentially generalizable? What are the classification systems? What do the design principles of the visualizations appear to be?

Tutorial exercises

1. Use the faceted classification system for television broadcasting units (Week 10) as a basis for the design of a cyberspace representation of the network's programming. Let your imagination go (assume very powerful but plausible technology). The system developed includes
   - Audience suitability ratings: (G, PG, MA, M)
     Expected viewership: (low, medium, high)
   - Product advertised: (various product categories, probably hierarchically organized)
   - Time slot: (Prime Time, Overnight, Midday, Morning, Saturday morning, Sunday morning)
   - Revenue-raising: (yes, no)
   - Spots: (Ad, Promo, Station logo, Public service announcement, Test pattern)
   - Ad type: (Single, Ad group, Ad reminder)
   - Target market: (Children, Adults, Family, Retired, Youth)
   - Program type: (News and Current Affairs, Sport, Information, Drama, Comedy, Music, Game, Variety)
   - N&CA type: (News, Sport, Weather, Current Affairs)
   - Info-type: (Documentary, Lifestyle, Talk show, Real life show, Religious, Open Learning)
   - Drama type: (Western, Police show, Soap opera, Hospital/ medical show, Australian outback show, Costume, Contemporary)
   - Language: (English, …)
   - Delivery: (Movie, Series, Mini-Series, Parts)
   - Comedy type: (Sitcom, Sketch, Amateur Video)
   - Medium: (Live actors, Animation)
2. Are any of the facets nominal? Does it make sense to see this classification system as a hierarchy? What would it look like visualized as a building or a city? What sort of content would you like to show? Can you show it using your visualization? How does this compare with the spatial representation of Question 1?

Open discussion question

 What are the implications of the solution of Question 1 for television broadcasting, given a sufficiently high bandwidth point-to-point communications network (upgrade to the existing Internet) and a powerful information visualization technology?

14 Archiving

Many information spaces are archives of one sort or another. Archives are an interesting example of the principles of this text. In addition, archiving poses its own set of problems.

What are archives?

Every human activity in a society involves pieces of information. In a literate society, these pieces of information are generally recorded, as written text, photographs, printed documents, films, e-mail messages, or on one of many other media.

Organizations operate on records. People keep records of their daily activities and their interactions with organizations. Some, generally organizations, are in the business of publishing information in one form or other.

Most of this information is very ephemeral. A bus ticket is important until the bus trip is completed. A cash register docket is important until we are sure we are not going to want to return the item purchased. A newspaper has value on the day it is published. Most of this information is discarded when no longer important. If we kept every bus ticket, register docket or newspaper we had ever used, we would drown in paper, not to mention the fire hazard.

Some information we keep. In particular, some records are evidence of rights or obligations. Our birth, marriage, divorce and so on are documented by official certificates, which we often need up to our deaths (which are also recorded) or even beyond (inheritances and so on). Our records of enrolment in university programmes and courses, grades and awards received can be important all our lives – we may need a transcript 30 years after graduation. Records of property ownership are central to the security of our homes. Towns in Europe often have records of privileges granted by the king in the 13th century. Every formal organization keeps minutes of decisions made by its governing body, which define the organization.

We also keep records of sentimental value. We may routinely discard bus tickets, but keep the tickets for our honeymoon trip. We may routinely discard cash register dockets but keep the one from our honeymoon hotel. We may routinely discard newspapers, but keep the one containing an article about the bravery award we won.

On a larger scale, records are of historical and cultural value. Detailed records of household expenses of a 19th century physician's family tell us a lot about how people lived in that time, place and social class. Newspaper records of concert programmes in the *Sydney Gazette* in 1828 tell us a great deal about the musical life of that time in Australia. A Sears, Roebuck catalogue for 1928 tells us a great deal about how people lived in rural America at that time.

The records we keep beyond their currency are, broadly speaking, archives.

How do we interpret an item of information?

Information comes in fragments. Take, for example, the photograph of Prudence Morgan milking a cow discussed in Chapter 4. The fragment of information is the photograph itself. Suppose you are rummaging in your grandparent's attic and come upon a photo of a woman milking a cow. You can tell the photo is old by its style (although we can make old-looking photos today) and by the costume and hairstyle of the woman (although people can dress in old style for re-enactments and so on). You take the photo to your grandmother, who tells you it is of Prudence Morgan, her grandmother's sister, taken at the dairy they had in Prairie Road the year her mother was born. You find that your grandmother's mother was born in 1884 from your grandmother's birth certificate. You find out from the family tree recorded in the old family Bible that Prudence was married to Elijah Morgan, who moved to Iowa from Vermont in 1876. You find a newspaper clipping in an old scrapbook that reports Elijah's retirement as President of the Acme Dairy in 1910, which says that the dairy specialized in milk from Holstein and Jersey cows. Another item in the scrapbook is a second prize at the State Fair in 1902 awarded to Prudence Morgan for butter made from Guernsey cream – the cow in the photo is thus very likely to be a house cow, not part of the dairy herd.

All this other information is needed to make sense of the photo. If the photo is to be kept in an organized archive, say the local Historical Society, all this contextual information must be gathered and recorded in an easily accessible form, probably in the catalogue entry for the photo. Ideally, the other items would also be in the collection, with the catalogue entries linked together in some way.

The photograph example is at a personal level. Organizational archives are often useful for cultural or historical studies. Personal records generally exist on pieces of paper which take considerable effort to keep, so only a tiny fraction of records end up as archives. Organizational information is much more structured and much more of it is held in information systems, so the decision to discard information is the decision that is made, rather than the decision to keep.

Aided by loyalty card programmes, many retail stores collect records of which person buys which products at which times. So my local supermarket knows that someone in my household bought a 300g jar of Kraft crunchy peanut butter on 1 July 2001. Someone in the early 22nd century might be interested in this piece of information.

Not, of course, because of any particular interest in an obscure middle-level academic of the late 20th and early 21st centuries. But they might be interested in the lifestyle of people with my sort of occupation, income level, education and residential area. To associate the record of purchase of the peanut butter with my social type, they need access to my loyalty card identification record, so that must be in the archive. The purchase record would be identified by personal identifier and a product barcode number, so to know that the purchase was of peanut butter, they need the store's product catalogue record, too. The product catalogue record links to a number of classifications of more and more general product types, which would be needed since the researcher's interest would be in a more general class of purchase than a 300g jar of Kraft peanut butter, crunchy. So these classification systems need to be in the archive.

Now it is unlikely that our early 22nd century researcher is particularly interested in 1 July 2001, or the Fairfield Gardens shopping centre in Brisbane, Australia. They would probably be interested in many areas across a period of perhaps 20 years, so ideally would like access to all the sales records for the store chain worldwide for the period of interest.

The research question might involve response of particular socio-economic groups to various forms of advertising for a type of product, so the store's advertising records would be of interest, including products featured, media and distribution. Links into the manufacturers' promotions might be indicated by special discounts and the like, revealed in the chain's purchasing records.

All this information is electronically acquired. So long as the data dictionaries and so on that determine the structure of the information are retained, it is possible to keep it indefinitely.

There is a large amount of it, though. To keep shopping details for 10,000 stores for 100 years would take on the order of 100 terabytes of data. Using present technology, one CD holds about 1 gigabyte, so this archive would be held on 100,000 CDs, which would occupy several cubic metres of space.

However valuable to 22nd century researchers, the cost of storing and maintaining the organization of these detailed records would be more than the store chain would consider it worth, and more than a library would be funded to support, given that this chain is only one of thousands with similar volumes of records. A decision would likely be made to archive the records in a highly aggregated form, reducing their volume by a factor of perhaps one million, so that the archives would fit on one CD. But this highly aggregated archive still needs the connections among the sales, product and purchasing information, otherwise it is uninterpretable.

Maintenance of context is particularly important in archives supporting rights and obligations. My descent from Charles II of Great Britain is only established if the chain of birth records extends from the present back to the 17th century in the various places my ancestors have lived. A certificate of property title only gives me the right to occupy a house if the land titles office can guarantee that no later title exists. A 13th century grant of privileges to a European town has force in the present only if an official archive can certify that the privileges weren't subsequently revoked or surrendered, perhaps during the 17th century religious wars.

In fact, the existence of reliable and authoritative archives of rights and obligations is central to modern civilization. It is hard to see how the real property system could exist without land title archives, nor legal systems without official archives of statutes and cases. These archives must be so reliable that it is almost unthinkable to question them.

So any item in an archive needs to be connected to many other items in order to retain enough context to be interpreted. Selection of records to retain or discard must take this need for context into account.

Design of an archive

We turn now to the issue of design of an archive as an information space. The first issue is to define the collection of documents – that is to say, which of the records supporting the operation of the organization or the life of the person we wish to keep.

To begin with, we clearly need to include any record of continuing obligations or privileges. As we have seen in the previous section, with these documents we need to include any other documents necessary to interpret them.

Beyond that, we need to include documents we think will have historical value. Here, the issue is one of cost and benefit. The benefits are almost by definition not known, since they will be reaped by people not yet known. The costs are known. They can be borne by either the body generating the documents or by some archival body. Governments often have agencies devoted to maintaining archives, with their own budgets. Specialized libraries or museums are often established by endowments or other funding sources. Historical societies keep archives, sometimes mainly with volunteer labour. As a result, generally speaking there will be a fixed budget for maintaining the archive. The selection of documents will therefore be made on the basis of an estimate of future value up to a budgeted cost, so that documents of lower value will be omitted. Furthermore, the budget allocated to archiving will generally be small compared with the costs involved in operating the information system supporting current activities.

So what has value? What has value is generally difficult to discuss in a general text of this kind. More so when the people to whom the collection will have value are often in the distant future. About the only clues we have are the kinds of questions present-day people ask of archives of the past. These seem to fall into three broad categories.

1. Major decisions made and the reasoning behind them. This would include minutes of Board or management committee meetings, briefing papers, consultants' reports and the like.
2. Summaries of the operation of the organization. These would be highly aggregated. Interest might be more in the system of classification used than in the totals themselves, so the classification systems themselves should be well documented.
3. Individual detail. As we have seen in the previous section, storing all the detail of the organization's activities is extremely impractical. The easiest thing to do from the point of view of current operations is to simply discard all detail when it is no longer needed. However, it would be a valuable addition to the archives if small samples were retained: a few customer transactions, a few product catalogues, a few advertisements, a few supplier invoices, a few interoffice memos, and so on. These could be selected at random. This is reminiscent of the use of randomly retrieved objects to understand the contents of classes as noted in the discussion of WebTOC in Chapter 13.

For organizations that publish information, the obvious thing to archive is the publications themselves. For any publication gone to press or put to air, however, there is generally a large body of drafts, notebooks, footage not used, background material and the like. In the past, some people have retained all this auxiliary material. There are many careers built on analysing the notebooks and drafts of James Joyce, Wittgenstein and Nietzsche. It would probably be valuable to include in the archive the complete material of a few publications. Again, because this is routinely discarded, there would need to be a conscious decision made to keep some of it.

Once the contents of a collection have been determined, the designer's attention turns to how to organize and index it. In Chapter 4 we saw that a key aspect of the design of the organization of an information space is to take into account the needs of the user population. However, in the case of archives, the user population is not known, much less what their information needs might be.

For this reason, archivists generally organize their collections on two bases:

- maintaining context;
- respecting time.

The former means maintaining the links among the individual items to aid in understanding the context. Respecting time means making it easy for the researcher to follow the time sequence of the origination of the documents in the collection. Whatever interest a researcher may have in a collection, documents produced earlier cannot have been made with the knowledge of the contents of documents produced later, so the historical sequence is both important and known by the archivist.

Since much of the most valuable information in archives is either private or sensitive for one reason or another, a central task of design of access to a collection is to preserve privacy and authorization of use. Many government organizations release the documents surrounding important decisions only 30 or 50 years after the events. In many countries, census forms are stored and released to the public only after 100 years. The archives must have been produced and organized long before their general release.

Obsolescence of media

Preserving the documents themselves is the most central task of the archivist. Organization and indexing can always be done if the documents themselves exist.

Information may be ethereal, but to exist it must be recorded on some physical medium. Physical media can be damaged or will deteriorate naturally. Ultimately, for information to be preserved it must be reproduced periodically. Certainly all the information we have from the ancient Greeks and Romans has been preserved by generations of copyists. Much has been lost. For example, we have none of the published work of even so influential a figure as Aristotle. The works we do have seem to be essentially lecture notes.

More recent material has been recorded on media with a long life. The original printed volumes of the works of Shakespeare or Newton have been preserved. However, the millions of copies of paperback books of say Isaac Asimov published in the second half of the 20th century will have disintegrated by 2050 due to acid deterioration of the paper on which they were printed. Prints of early films have been lost due to deterioration of their nitrate-based film stock.

But the problem is much more severe in the present age of high-technology media.

In the early 1980s, the author was responsible for a suite of accounting software worth about US$1 million. This software was archived on a removable 25-megabyte disk backed up on a tape cartridge and 8 inch floppy disks. It is unlikely that any of the media have survived 20 years. Even if they had survived, the computer systems needed to read the disks and tapes no longer exist. Even if one could find a hardware platform that had the necessary peripheral devices and driver software, the information would be difficult to recover without the operating systems, wordprocessors, compilers and other elements of the programming environment with which the software was created.

Eight-inch floppy disks were replaced by 5 1/4-inch floppy disks, which were replaced with 3 1/2-inch floppy disks, originally with 250 kilobytes capacity, then 500 kilobytes, then 1 megabyte. At the time of writing, floppies are being phased out altogether in favour of CD-ROMs. Betamax videotapes and 8-track audiotapes are unplayable today – even if a reader could be resurrected, the media have deteriorated. It looks as if VHS videotapes will be replaced with DVD. A similar story can be told in sound recording, data recording, video, film and the Internet.

So an urgent issue in planning and maintaining an archive is the preservation of the media on which the information is recorded. We must reproduce the information at a frequency determined by the lifetime of the media. Either we must maintain archives of the hardware and software platforms supporting the media, or we must copy the information to new media and platforms, which involves translating the software formats as well as copying the files.

Even the older, more permanent media are affected by this problem. It is very common for organizations responsible for paper-based archives to digitize their collections. Digitization makes access to the material much easier, but introduces the media and platform preservation problem.

Then there are virtual documents. Many important decisions are taken on the basis of documents dynamically produced from a repository on the basis of a query. What do we do about these?

All these issues are under active consideration by the people responsible for the design and maintenance of archives. No general solutions have emerged as yet.

Key concepts

Archives must be organized to maintain the **context** and **temporal sequence** of the documents they contain. Planning for archives must take into account **deterioration of media** and **technological obsolescence** of hardware and software platforms.

Further reading

There are many books on archiving principles. This xhapter was prepared with the assistance of Ellis (1993). Discussion of media deterioration and technological obsolescence is becoming more prominent in the professional literature.

Formative exercise

Look into an archive in an area with which you are familiar.

1. What types of documents are included?
2. How is the archive organized to maintain context?
3. How is it organized to maintain the temporal sequence of the documents?

4. Are there any other principles of organization?
5. What media are the documents recorded on?
6. What hardware and software platforms are necessary to make use of the media?
7. What policies and procedures does the archive have to conserve the information?
8. How does the archive deal with technological obsolescence?

Tutorial exercises

 There is a large amount of information associated with a university course using this text. Besides the teaching material, the lecturer keeps records of all e-mails in and out, all contributions to e-mail tutorials, name and student number and degree programme for each student, marks for assignments, indications of late submission, special agreements, grade, topic for each student's major assignment, e-mail addresses, records of tutorial group membership, records of students who have enrolled but later withdrew.

1. How much of this should I archive? Take into consideration
   - Who needs to see what for what time period?
   - Is some information archived by someone else?
   - Media and platform issues.
   - Privacy.
2. What sorts of questions could be answered from such an archive in, say, 10 years?
3. Would it be advantageous to have available electronically and what sorts of things not?

Open discussion question

It is technically possible for all of the information you need for your university education to be published electronically (except of course what you get from face-to-face contact with lecturers, tutors and other students, which is extremely important). What sorts of things would you like to see available in some electronic form, and what not? Take a personal perspective: what would you prefer and make good use of? Are there any feasible changes to either technology or the organization of the information that would make any difference?

15 Quality

Quality is an important dimension in publishing information or in assessing an information source. The most significant quality issues are where information brushes with the law.

Quality of content

A published information space includes three aspects: the content, its structure and its quality. Content is specific to the project, so can be treated in a general text of this kind only in very general ways. Most of this text is in fact about the structure of the space. In this final chapter we look at the issue of quality.

Quality has many aspects. There is a technical dimension – the response time or the number of users that can simultaneously access the source at an acceptable response time. There is a media dimension – does the medium chosen permit sufficient presentation and access? A user interface dimension – how easy is the source to use? A business dimension – is the source profitable? These aspects are all important to the overall success of the project producing the space, but are outside the scope of this text.

We will concentrate on four aspects of quality: authoritativeness, completeness, timeliness and assimilability.

Authoritativeness means that the information in the space is the best available. A weather report published by a government Bureau of Meteorology for a particular area is likely to be the best weather report for that area. Sporting event results from the Olympic Games are likely to be most authoritative if they come from the official Olympic Web site. The *Encyclopedia Brittanica* has long had a reputation as an authoritative source on the issues it covers. These sources get their authority from respectively the most comprehensive data collection, best qualified scientists and most powerful analytic tools (meteorology); from being closest to the source (Olympics), or from painstaking research by the best minds (encyclopedia).

One dimension of authoritativeness for certain kinds of publication, in particular financial data, involves certification by an auditing body. Auditors are used to working with a fixed set of documents that, once certified, are almost impossible to change. However, a document published on the Internet can be changed at will. The following newspaper clipping shows that the auditing profession is worried about this.

Auditors push to tame unruly Web

COMPANY auditors, concerned about putting their names to online information, have proposed tighter rules to govern the use of supposedly audited financial data on their clients' Web sites. The audit profession is increasingly worried about how far their audit assurance statements should apply to financial data originally drawn up in paper form and then repeated – but possibly changed or updated – online.

As a precautionary step, the Auditing and Assurance Standards Board has produced a "guidance statement" which is directed at auditors but also dispenses advice to management. The statement stresses that a normal audit statement does not cover the quality and security of Web site information (*The Australian*, 20 January 2000, p. 20).

Completeness means that if a user goes to a site looking for information in the general topic areas covered by the space, they are very likely to find it. Amazon.com and other major boor retailer sites, for example, have access to the indexes of books in print, so if a book is in print, they know about it. The Olympic site has complete information on events, competitors and results. An official university catalogue has complete information about the programmes and courses offered.

Timeliness has to do with information that dates. A weather report is useful for only a few hours before it is superseded. The official meteorological reports are where the forecasts are published, so would be the most timely. Company financial reports are published quarterly in many countries – the performance can change markedly from quarter to quarter. The official announcements are the most timely (getting in before the official announcement is what is called "insider trading" and is against the law in most places). In an organization, people enter and leave and move around, so their telephone numbers are entered, deleted or changed often. The PABX is the most timely source of telephone numbers.

Assimilability is a different sort of property from the others. Authoritativeness, completeness and timeliness are all properties of the source itself. We can talk about these properties in absolute terms. Assimilability refers to the ability of a user of some information to take it in to the point of being able to make use of it. Information economists recognize that the usefulness of a body of information to a consumer can only be assessed after the consumer has taken it in and understood it. Hence the slogan *useless information is worth less than nothing.* A given piece of information can be very easy for one consumer to assimilate and very difficult for another.

Consider information on pharmaceuticals, for example. There is a large amount of information on the Internet about various therapeutic drugs. Site A is a search engine returning all available high-quality literature on a nominated drug, while site B is a very sparse table giving only the dosages, contraindications and major side effects, with a link to the most recent literature. If I as a patient have been prescribed a drug, I may look at site A for more information about it. Site B would be incomprehensible to me as it is couched in a very condensed, highly technical language. In other words, site A is much more assimilable to me.

However, a practising physician often needs specific information about drugs in the process of a consultation. Site A would give far more information than is needed, and it would take the physician a long time to sift through and get what is wanted. Site B is targeted to practitioners in this situation, so gives exactly what is required in a condensed, highly technical language that the physician understands through long training. So for this consumer, site B is far more assimilable.

A high-quality source of valuable information is in an excellent position to benefit from Internet access, so long as its contents are difficult to copy. Alternative sites are expensive to establish and it is very expensive to develop a reputation for quality. The high quality site can charge a higher fee than a low-quality site so long as the consumers understand what quality means for that service, because the cost of assessing the quality of an unknown site can be very high. This cost cannot necessarily be bypassed by relying on others' judgments due to differences among what people find easily assimilable.

Legal issues – overview

One thing that can happen in publishing information is that one can run afoul of the law. This can result in fines, or shutting the publication down, or even jail terms for the publisher. A publication where one of these happens can hardly be considered high quality, even though it may satisfy the more technical criteria of the previous section. Understanding what these issues are is important.

Several kinds of things can go wrong. One can break the criminal law, either as a personal offence or publishing forbidden material. One can be sued under the civil law for libel, or for privacy violation, or for infringement of intellectual property.

In terms of the previous section, covering these topics in a part of a small, final chapter in a textbook of this kind can hardly be considered a high-quality source of information. Furthermore, in the Internet space these kinds of issues are in a state of considerable flux. It is simply not known what the rules are in many cases.

Therefore, what this chapter will do is to describe briefly and generally what the issues are, then will discuss some recent cases which bring these issues into focus. The reader is recommended to consult experts or to look into specific authoritative literature for reliable guidance, where such exists.

Law is a collection of norms, rules and regulations governing behaviour in societies. Specific behaviour is regulated by judicial bodies set up by government and similar agencies. Any judicial body has a *jurisdiction*, which restricts its activities to certain classes of behaviour in certain geographical areas. The laws of the State of Queensland, Australia, do not apply to people residing in Seattle or Vladivostok. To sue someone for infringement of copyright one generally uses a different court than one uses to charge someone with publishing child pornography, a criminal offence in many places.

The *criminal law* forbids a wide variety of behaviours, some of which are information-related. These include harassment, fraud and theft among many others. Classed under *censorship* are laws forbidding publication of certain kinds of materials. These laws differ greatly from jurisdiction to jurisdiction.

Libel is the publication of material which unjustifiably or maliciously harms the reputation of a person or sometimes an organization. *Privacy* laws prevent the publication or distribution of information of a private nature gathered about a person or organization.

Intellectual property is the most difficult of all. The primary form of intellectual property involving publishing is copyright. The originator of a work has the legal right to control its publication and to have a share in any revenue generated from publication. The originators very often assign these rights to publishers in return for royalty payments. Unauthorized copying is a violation of copyright, and whoever does so can be sued.

Intellectual property is also protected by patents and by trade secrets. Various technical means exist to control distribution and prevent copying, including encryption.

The remainder of this chapter is a collection of recent events which cast some light on some of these issues and how the norms and regulations are changing as a result of the Internet.

Crime

Internet threats of rape earn fine

A MAN who threatened to gang rape a woman he had never met and sell a video of the assault on the Internet pleaded guilty yesterday in a Brisbane court. Jevon Antony Rowan Ingram, 27, pleaded guilty to using a carriage service for offensive behaviour in one of the first cases of Internet stalking in Queensland courts (Brisbane, Queensland, Australia *Courier-Mail* 20 May 2000, p. 7).

Making threats like this is a criminal offence; however, the threats were communicated to the victim. In this case, the threats were sent via the Internet. So far as that goes, this case is entirely unremarkable. The interesting feature is what it says about jurisdictions.

It happens that in this case both the perpetrator and victim resided in the city of Brisbane, in the state of Queensland, in the country Australia. What would have happened had the perpetrator lived in Vladivostok? Or suppose the perpetrator lived in Brisbane, but the victim lived in Seattle. It is not obvious what court would have jurisdiction, and how the case could proceed.

Censorship

Censor's axe falls on 22 Web sites

JUST 22 Web sites have been issued with takedown notices since the [Australian] Federal Government's Internet regulation laws came into effect at the start of the year.

An ABA [Australian Broadcasting Authority] spokeswoman said the takedown notices applied to sites that were refused classification (RC) under the Office of Film and Literature Classification's (OFLC) R, X or RC ratings. The sites that were refused classification were beyond the X-rating, the spokesperson said. "The majority of sites were related to sexual material," she said (*The Australian*, 14 March 2000, p. 43).

US court rules out pro-life Web site

IN a landmark ruling on free speech and the Internet, a US court has found that a Web site splattered with simulated foetal blood and containing the names and addresses of abortion doctors amounted to a death threat against the physicians. The verdict, which was a blow to anti-abortion groups, has highlighted the limits of the constitutional protection given to speech on the Web. It was also seen as an important legal test of aggressive tactics employed by militant antiabortion groups in the US (*The Australian*, 4 February 1999, p. 8).

Many countries have begun to successfully censor sites containing pornography or other objectionable material. But what about if the site is not located in the country concerned? Some courts in some countries are attempting to censor information originating elsewhere. It remains to be seen what effect these rulings will have.

Burying a Nazi piece of work

America's leading Internet portal has been ordered to block French users from viewing and obtaining Nazi material in a ruling that threatens freedom of the web. In a landmark ruling that will have implications for the operators of websites everywhere, a French court has ordered that US Internet giant Yahoo must prevent users in France from having access to Nazi memorabilia offered on its US auction site, or face heavy fines. Even though it is unclear whether the ruling is enforceable, especially through the American courts, the case is the first serious attempt by one country to impose its laws on a technology that seemed to know no international borders (*The Australian*, 24 November 2000, p. 28).

Libel

Libel cases are generally taken against publishers in the mass media. The Internet by itself is not much different from television or newspapers as a publishing medium. However, the following report shows that the Internet changes the problem.

In cyberspace, libel can be writ large

THE real world is catching up on the Internet, with what is thought to be one of Australia's first defamation actions based on information posted on the Net. Brisbane-based First Australian Building Society is suing the Heritage Building Society over a press release posted on its Web site on Tuesday (*The Australian*, 19 February 2000, p. 35).

The problem is that in the mass media, a company may issue a press release and a news medium may publish an article based on the release, but the news medium takes responsibility for libel. They are required to cross-check sources, and do other things to avoid libelous statements. Publishing the press release on your own Web site means that you take the responsibility, so open yourself to libel suits.

Privacy

Privacy has long been an issue in industrial societies. The Internet has simply changed the medium that the problem occurs in. Scott McNealy, of Sun Microsystems, has been quoted as saying, "There is no privacy. Get over it". At least at the time of writing, that view seems to be in a minority.

Many organizations collect information about people and other organizations as a by-product of their operation. Bank records, credit records, taxation records, motor vehicle records, records of goods purchased (if you have a retail loyalty card, then everything you have ever bought at participating stores is recorded against your identifier) are a few examples. In many jurisdictions it is forbidden for this information to be accessed by unauthorized people, and it is forbidden to use that information for other purposes without the consent of the people involved.

The Internet exacerbates the problem by enabling online shopping or purchase, which almost by definition requires identifying the purchaser, so that in effect you have a loyalty card with every shop you buy at. Many of these online shops are much smaller and less experienced that the large chains that normally operate loyalty card schemes.

Many organizations gather information about people or other organizations covertly – without the knowledge of the people being monitored. These in the past have generally been government agencies, including police and intelligence agencies, operating under more or less strict government guidelines. Commercial market intelligence companies collect and collate publicly available information. Again, there are restrictions on unauthorized use, and restrictions on what this information can be used for.

The Internet changes covert monitoring by making aspects of it much easier. Many sites leave small pieces of software called *cookies* on users' machines that are capable of monitoring what Web sites the user's machine accesses, and reporting summaries to the original site. Controlling who gets to do this, with what kind of consent, and what the information can be used for, are all in contention at the time of writing.

Data gatherers caught with their hands in the cookie jar

Recent privacy incidents challenge fundamental Internet business models – many of which are premised on cost-free information gathering, information reselling and close ad targeting. The first offenders were several health information sites, which used information covertly gleaned from consumer viewing patterns to sell ad slots. In February, the US Federal Trade Commission announced that it would scrutinise several online health players.

A bigger issue has more recently arisen around Doubleclick, the Internet's largest marketing player. It is facing several class action lawsuits, two investigations by state regulators and a probe from the Federal Trade Commission over its privacy practices. It has been alleged that the firm misleads consumers over both the data it collects on them and the uses to which that data is put (*The Australian*, 9 March 2000, p. 24).

Finally, the Internet makes it much easier to gather, collate, and publish information which is in the public record but which is otherwise difficult to access. For example:

Net makes privacy law a pressing issue

PRIVACY is one of the most important issues facing the Internet community. Last week the launch of a site called CrimeNet emphasised this more than ever. The site provides "a complete information service on criminal records, stolen property, missing persons, wanted persons, con artists and unsolved crimes", according to the front page of the site (www.crimenet.com.au/).

The people behind the site collect information from newspapers and other media sources and enter it into a database. The site then levies a small charge for people wanting to search some parts of the database. The people behind the site argue that because the information is already public, and the courts are open, they are doing society a service by bringing all this information together in one place (Jeremy Horey, in *The Australian*, 9 May 2000, p. 8 Cutting Edge).

You may have done something which was reported in a newspaper a long time ago in a different city. The report may have been inaccurate, but no correction was ever published. Up to now, this has not been a problem for you, since only a very determined person (perhaps one of the intelligence agencies) would find the information. Now everyone can know in seconds. What difference does this make to you? These sorts of publications have created enormous controversy.

Intellectual property

Intellectual property is probably the biggest issue involving the Internet at the time of writing, since intellectual property is the basis for a very large fraction of the world's economy. Copyright and other protection of such property is well established. However, the Internet adds some interesting twists.

In particular, it is easy to include someone else's intellectual property in a product of your own, more or less masking the original. A case involving shareware was reported in *IEEE Computer* in 1996.

Shareware, as many readers would know, includes programs which can be downloaded free and used for a limited trial period. If someone wants to continue using the product, they are required to pay a generally small sum to the developer. This payment is not generally enforced, but depends on the honesty of the user. The terms of a shareware distribution are generally included in a licence agreement which accompanies the downloaded software.

The case reported involves a software product at one time widely used for connection to the Internet, called Trumpet Winsock. As part of a promotion, an Australian ISP distributed floppy disks containing software that enabled a user to connect to their service. The disk included Trumpet Winsock among other software, but packaged so that its parameters were pre-set. There was also an error in the parameter file. This distribution was done without permission.

Trumpet sued the ISP for copyright violation. The ISP maintained that they were simply distributing shareware, while Trumpet argued that Winsock was so packaged that the user would not be aware that they had an obligation to pay Trumpet for its use.

In what is claimed to be the first case anywhere in the world to rule on copyright protection for shareware, a court in Tasmania, Australia, ruled in favour of Trumpet. The ruling allows third parties to distribute shareware, but only in unmodified form and in such a way that the shareware licence agreement is not obscured.

A related problem occurs on the Web, where it is easy for one service to put a wrapper around another so as to include the second service as an element of the first.

Newspaper job site Monstered

CareerPath [a recruiting site] came up with a device called Resume Scout which, for $200 a month, allows recruiters to search multiple paid and free resume databases simultaneously – including Monster.com. A recruiter who is already a paid subscriber to Monster.com can use Resume Scout to access its posted resumes without ever going directly to the site.

Monster isn't taking too kindly to this encroachment. Myron Olesnyckyj, Monster's in-house attorney, sent CareerPath a cease-and-desist letter, warning the company that Monster.com's Terms of Use prohibit the use of any "browser, spider, robot, avatar or intelligent agent" to search its Website (*The Australian*, 5 May 2000, p. 30).

The problem here was similar to the Winsock case in that CareerPath's link to Monster was only able to be used by people registered with Monster, so CareerPath did not steal Monster's service. The issue was that Monster lost its identity, and also any advertising on the Monster site was not visible to those who accessed it via CareerPath. The only way to do this sort of thing in the physical world would be to find a popular store and build a façade over the façade of that store, with your name over your door which led immediately to the door of the original store – not very practical in the real world but easy on the Web.

Napster

Probably the most economically significant issue in intellectual property in the year or two prior to the time of writing is the Napster saga, because it affects the very foundation of the recorded music industry.

Music is pure information. In digital form, a piece of music is a string of zeroes and ones. Historically, music has been distributed in the form of physical objects (records, CDs, tapes), which are warehoused, held in retail inventory and sold in much the same way as clothing, groceries or mobile phones. However, the bit strings can be stored in a computer system and re-played using a computer system equipped with a digital-to-analogue converter and speakers, now quite common. Further, these bit strings can be compressed using procedures like MPEG, thereby reducing the amount of space needed to store them and the cost of transmitting them over the Internet.

Copying of music has always been a problem for the industry. However, recording from a broadcast generally reduces quality significantly, and copying onto one of the physical media requires some effort and more than minimal equipment, so the companies have generally not worried too much about copying for personal use. Their intellectual property protection efforts have been concentrated on organized commercial piracy.

However, once the music tracks are stored on people's computer systems, copying them and sending them on becomes extremely simple. Moving beyond simply copying tracks to friends, or copying from home to work, in the late 1990s several so-called *peer-to-peer* communication protocols were introduced permitting such copying on a large scale.

The most prominent of these protocols was Napster. If I know that person X has a particular piece of music on their disk, the protocol allows me to copy it from their disk to mine. There is no central repository of music tracks, just the ability to copy from one user to another (hence the term peer-to-peer). When millions of people started copying material from millions of other people, much of it copyright, the recording companies got upset and sued Napster.

Napster argued that it was just a communications facilitator, and what the users did with its protocols was the users' business, not Napster's responsibility. However, Napster did maintain a directory of available tracks (the obvious way for me to find out who has the track I am interested in). At the time of writing, Napster has agreed with several major recording companies to charge a fee for downloading their material sufficient to pay royalties. So Napster is moving from a free service to a subscription service. However, non-copyright material can still be copied free.

This settlement is by no means the end of the story. For one thing, the record companies need to be able to identify which tracks stored as files on millions of computers contain material whose copyright they own. This is proving difficult, since the mechanism used is lists of file names, and people can easily change the names of the files stored on their systems.

More seriously, there are several other peer-to-peer protocols which do not rely on a centralized directory of content, including Gnutella. These protocols, naturally, are increasing in popularity. They are much more difficult for the record companies to sue, since they have no centre. They are much more like sharing among friends on a very large scale. One person has in effect 100,000 friends instead of 10.

One development that may be significant is that the record companies have begun distributing music tracks over the Web using Napster and other services at a much lower price per track than one would pay if one purchased the physical record product. This makes the cost to the user of a legally obtained track much lower. On the other hand, the free services suffer from lack of quality control, for example:

Can New Technology Revolutionise the Net?

For example, yesterday I was playing around with Gnutella and typed in the word "dinosaur" (I have a 2-1/2-year-old at home with a major Stegosaurus fetish). Half the results led to a deliberately mislabeled HTML page that turned out to be an ad for porn videos. No dinosaur fossils were in evidence, though there was no shortage of skin over bone. Or consider the consequences of an increasingly common prank called "Napster bombing." I've heard of cases in which an MP3 file purporting to be from teenybopper Britney Spears turned out to be a track from rude hip-hopper (and Spears critic) Eminem. How do you know files you're downloading through Gnutella aren't similarly mislabeled, or worse, harboring nasty viruses? If the recent "resume" for Janet Simons could turn out to be a hard-drive-ravaging worm, imagine what might be lurking in other innocent-sounding files. I'm describing a worst-case scenario, but decentralised content, though tempting in an egalitarian sort of way, is just plain dangerous (Steve Fox, Editor, CNET Online 1 June 2000).

We come full circle to the issue of quality with which we began this chapter. Quality of information is worth something. So long as the cost of information from a high-quality source is not much more than the cost of ascertaining the quality of an unknown source, the quality source can charge a premium price. This may be the future of not only the music industry, but also DVD distribution, games, electronic books, and scientific journals.

Key concepts

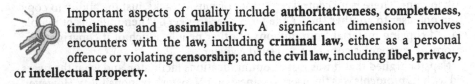

Important aspects of quality include **authoritativeness, completeness, timeliness** and **assimilability**. A significant dimension involves encounters with the law, including **criminal law**, either as a personal offence or violating **censorship**; and the **civil law**, including **libel**, **privacy**, or **intellectual** property.

Further reading

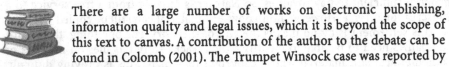

There are a large number of works on electronic publishing, information quality and legal issues, which it is beyond the scope of this text to canvas. A contribution of the author to the debate can be found in Colomb (2001). The Trumpet Winsock case was reported by Cifuentes and Fitzgerald (1996).

Formative exercise

Consider your favourite information sources. How do you ascertain their quality compared with alternative sources which you know about but do not use? How do they protect their intellectual property?

Tutorial exercises

Suppose one wanted to publish the material used in a course based on this text. Which of the issues mentioned in the chapter are relevant, and how might they be addressed?

1. What? (types of documents).
2. How? (HTML, PDF, Postscript, Database, etc.).
3. Quality and quality assurance.
   - authoritativeness, completeness, timeliness and assimilability.
4. Legal issues
   - libel
   - copyright
   - privacy.
5. Agreements for stability of third party services.

Open discussion question

Consider an area of publishing with which you are familiar. What are the implications of Internet publishing for that industry? Take into account the full spectrum of relevant issues, including feasibility, acceptability to the consumer, quality, legal issues, impact on existing channels, cost, and so on.

Major assignment

This text was intended to introduce the reader to a collection of concepts which form a critical vocabulary for the design and criticism of information publications. This exercise is designed to give the student an opportunity to use a wide range of these concepts while performing a specific concrete task.

The task

Write a 2000–2500 word report which does one of the following:

- Critically evaluate a substantial source of information.
- Compare two source of information.
- Design a substantial source of information.

Take into account the following:

- The service that the site is intended to provide. Who are its users intended to be? What information should be provided to them in order to satisfy their information requirements? What will this information be used for? How would it ideally be accessed? This aspect is irrespective of what the service actually does, and is intended to be a basis for criticizing the implementation.
- To what extent does the actual system have the information and access methods you thought useful? If some features are missing, why in your opinion are they missing? If you are designing a system, why might some features be difficult to implement?
- How (whether) and to what end it uses classification, keywords and hypertext links. If one or more of these are not used, explain why they are not used, and whether it makes sense to modify the system to use them.
- How does the medium in which the system is implemented help or hinder its effectiveness? Illustrate your claims with concrete scenarios. Show at least one help and one hindrance.
- Use a vocabulary derived from your reading, supported by citations if outside the required reading. (For example, if classification is used, is the system enumerative, faceted or what? How are keywords obtained? etc.) Note that the system does not need to be computerized at present. If it is not, you should include a discussion of the feasibility of computerization and implications for the information source if it were to be computerized.

Assessment is intended to test the student's command of the vocabulary in specific situations. However, a major part of the credit would be assigned to satisfactory completion of the specific tasks in the specification. In general:

- A satisfactory command is to be able to use concepts appropriate to a situation.
- A superior command is to be able to understand the relative importance of the various aspects of the conceptual structure (for example which aspects are central to the problem, which are more peripheral, how the various aspects interact).
- An excellent command is to be able to see the strengths and weaknesses of the conceptual structure as revealed by application to a specific situation, demonstrated by going beyond the material in some way, for example insight into the implications of the problem and its solution for information science itself.

Suggestions for the report

Some students put considerable work into their reports for the topics like this, but do not reap marks commensurate with the effort they put in, as their effort was devoted to aspects of the problem which are outside the assessment. It takes considerable skill to write to the point, a skill which is much more general than this course. The following may help you gain the marks your effort deserves, and help you in the future.

The purpose of the assignment is to show understanding of the material from the course. Understanding the material from the course means that the student will learn a specialized technical vocabulary and its use in design and criticism of design of information spaces. The particular task in the assignment is designed to give you an opportunity to demonstrate that you know what the vocabulary is and how to use it.

Any information source can be viewed from a number of different aspects. The relevant aspect is structural, from an information science perspective. However, it can be looked at from a user interface, business, technological, content, or marketing perspective, among probably others. Issues from these aspects are relevant only to the extent that they have impact on the information science aspect. For example:

- we don't care if the user interface is any good, but user interface considerations may influence the choice of classifications – e.g. to achieve an aesthetically pleasing presentation certain objects are grouped which might not otherwise be grouped;
- we don't care whether the source is profitable, but business considerations may influence which information science based facilities are provided – e.g. a festival program may not provide a good index so that people will have to browse the program therefore exposing themselves to more of the advertisements;
- we don't care how the source is implemented, but technological considerations may influence which information science based facilities are provided – e.g. a paper-based source can have a large and high-resolution map covering an entire information space, while a computer-based solution must make do with low resolution and zooming, so needs a classification system which the paper-based solution does not;

- we don't care whether the content of the source is interesting, correct, or where it comes from, but the nature of the content makes a big difference to the information science based facilities provided – e.g. a compendium of several reference books will typically not be able to have a uniform index across all content whereas a large encyclopedia managed by a single organization can;
- we don't care who the site is aimed at, but the target audience has a big impact on the organization of the space – e.g. a space aimed at a varied population of casual users may rely mainly on a classification system represented as a menu structure, while a space aimed at a specialized group of frequent users may rely more on a keyword index.

The skill is to stop before you do a large amount of work on some aspect of the problem, and assess how the results you will get can have an impact on your terms of reference. If it will have little impact, don't pursue it. When you finish your first draft, examine each section to see how it contributes. You always have a limited word budget, so if in doubt, leave it out. Also, in general, hyperbole is a waste of words. If someone wants a report on a particular topic, they don't need to be told why the topic is a good idea.

US Department of Labor Standard Industrial Classification division structure, 1987 edition

A. Division A: Agriculture, Forestry, and Fishing
   - Major Group 01: Agricultural Production Crops
   - Major Group 02: Agricultural Production Livestock and Animal Specialties
   - Major Group 07: Agricultural Services
   - Major Group 08: Forestry (see below for further expansion)
   - Major Group 09: Fishing, Hunting, and Trapping
B. Division B: Mining
   - Major Group 10: Metal Mining
   - Major Group 12: Coal Mining
   - Major Group 13: Oil and Gas Extraction
   - Major Group 14: Mining and Quarrying of Nonmetallic Minerals, Except Fuels
C. Division C: Construction
   - Major Group 15: Building Construction General Contractors and Operative Builders
   - Major Group 16: Heavy Construction Other Than Building Construction Contractors
   - Major Group 17: Construction Special Trade Contractors
D. Division D: Manufacturing
   - Major Group 20: Food and Kindred Products
   - Major Group 21: Tobacco Products
   - Major Group 22: Textile Mill Products
   - Major Group 23: Apparel and other Finished Products Made From Fabrics and Similar Materials
   - Major Group 24: Lumber and Wood Products, Except Furniture
   - Major Group 25: Furniture and Fixtures
   - Major Group 26: Paper and Allied Products
   - Major Group 27: Printing, Publishing, and Allied Industries
   - Major Group 28: Chemicals and Allied Products
   - Major Group 29: Petroleum Refining and Related Industries

- Major Group 30: Rubber and Miscellaneous Plastics Products
- Major Group 31: Leather and Leather Products
- Major Group 32: Stone, Clay, Glass, and Concrete Products
- Major Group 33: Primary Metal Industries
- Major Group 34: Fabricated Metal Products, Except Machinery and Transportation Equipment
- Major Group 35: Industrial and Commercial Machinery and Computer Equipment
- Major Group 36: Electronic and Other Electrical Equipment and Components, Except Computer Equipment
- Major Group 37: Transportation Equipment
- Major Group 38: Measuring, Analyzing, and Controlling Instruments; Photographic, Medical and Optical Goods; Watches and Clocks
- Major Group 39: Miscellaneous Manufacturing Industries

E. Division E: Transportation, Communications, Electric, Gas, and Sanitary Services
- Major Group 40: Railroad Transportation
- Major Group 41: Local and Suburban Transit and Interurban Highway Passenger Transportation
- Major Group 42: Motor Freight Transportation and Warehousing
- Major Group 43: United States Postal Service
- Major Group 44: Water Transportation
- Major Group 45: Transportation by Air
- Major Group 46: Pipelines, Except Natural Gas
- Major Group 47: Transportation Services
- Major Group 48: Communications
- Major Group 49: Electric, Gas, and Sanitary Services

F. Division F: Wholesale Trade
- Major Group 50: Wholesale Trade-durable Goods
- Major Group 5 1: Wholesale Trade-non-durable Goods

G. Division G: Retail Trade
- Major Group 52: Building Materials, Hardware, Garden Supply, and Mobile Home Dealers
- Major Group 53: General Merchandise Stores
- Major Group 54: Food Stores
- Major Group 55: Automotive Dealers and Gasoline Service Stations
- Major Group 56: Apparel and Accessory Stores
- Major Group 57: Home Furniture, Furnishings, and Equipment Stores
- Major Group 58: Eating and Drinking Places
- Major Group 59: Miscellaneous Retail

H. Division H: Finance, Insurance, and Real Estate
- Major Group 60: Depository Institutions
- Major Group 61: Non-depository Credit Institutions

- Major Group 62: Security and Commodity Brokers, Dealers, Exchanges, and Services
- Major Group 63: Insurance Carriers
- Major Group 64: Insurance Agents, Brokers, and Service
- Major Group 65: Real Estate
- Major Group 67: Holding and Other Investment Offices

I. Division I: Services

- Major Group 70: Hotels, Rooming Houses, Camps, and Other Lodging Places
- Major Group 72: Personal Services
- Major Group 73: Business Services (see below for further expansion)
- Major Group 75: Automotive Repair, Services, and Parking
- Major Group 76: Miscellaneous Repair Services
- Major Group 78: Motion Pictures
- Major Group 79: Amusement and Recreation Services
- Major Group 80: Health Services
- Major Group 8 1: Legal Services
- Major Group 82: Educational Services
- Major Group 83: Social Services
- Major Group 84: Museums, Art Galleries, and Botanical and Zoological Gardens
- Major Group 86: Membership Organizations
- Major Group 87: Engineering, Accounting, Research, Management, and Related Services
- Major Group 88: Private Households

J. Division J: Public Administration

- Major Group 91: Executive, Legislative, and General Government, Except Finance
- Major Group 92: Justice, Public Order, and Safety
- Major Group 93: Public Finance, Taxation, and Monetary Policy
- Major Group 94: Administration of Human Resource Programs
- Major Group 95: Administration of Environmental Quality and Housing Programs
- Major Group 96: Administration of Economic Programs
- Major Group 97: National Security and International Affairs
- Major Group 99: Nonclassifiable Establishments

SIC Major Group 08: Forestry

Major Group Structure

This major group includes establishments primarily engaged in the operation of timber tracts, tree farms, forest nurseries, and related activities such as reforestation services and the gathering of gums, barks, balsam needles, maple sap, Spanish moss, and other forest products.

Industry Group 08 1: Timber Tracts
   0811 Timber Tracts
Industry Group 083: Forest Nurseries and Gathering of Forest
   0831 Forest Nurseries and Gathering of Forest Products
Industry Group 085: Forestry Services
   0851 Forestry Services

SIC Major Group 73: Business Services

Major Group Structure

This major group includes establishments primarily engaged in rendering services, not elsewhere classified, to business establishments on a contract or fee basis, such as advertising, credit reporting, collection of claims, mailing, reproduction, stenographic, news syndicates, computer programming, photocopying, duplicating, data processing, services to buildings, and help supply services. Establishments primarily engaged in providing engineering, accounting, research, management, and related services are classified in Major Group 87. Establishments which provide specialized services closely allied to activities covered in other divisions are classified in such divisions.

Industry Group 731: Advertising
   7311 Advertising Agencies
   7312 Outdoor Advertising Services
   7313 Radio, Television, And Publishers' Advertising Representatives
   7319 Advertising, Not Elsewhere Classified
Industry Group 732: Consumer Credit Reporting Agencies, Mercantile
   7322 Adjustment and Collection Services
   7323 Credit Reporting Services
Industry Group 733: Mailing, Reproduction, Commercial Art And
   7331 Direct Mail Advertising Services
   7334 Photocopying and Duplicating Services
   7335 Commercial Photography
   7336 Commercial Art and Graphic Design
   7338 Secretarial and Court Reporting Services
Industry Group 734: Services to Dwellings and Other Buildings
   7342 Disinfecting and Pest Control Services
   7349 Building Cleaning and Maintenance Services, Not Elsewhere Classified
Industry Group 735: Miscellaneous Equipment Rental and Leasing
   7352 Medical Equipment Rental and Leasing
   7353 Heavy Construction Equipment Rental and Leasing
   7359 Equipment Rental and Leasing, Not Elsewhere Classified
Industry Group 736: Personnel Supply Services
   7361 Employment Agencies
   7363 Help Supply Services
Industry Group 737: Computer Programming, Data Processing, And
   7371 Computer Programming Services
   7372 Prepackaged Software
   7373 Computer Integrated Systems Design
   7374 Computer Processing and Data Preparation and Processing Services

7375 Information Retrieval Services
7376 Computer Facilities Management Services
7377 Computer Rental and Leasing
7378 Computer Maintenance and Repair
7379 Computer Related Services, Not Elsewhere Classified
Industry Group 738: Miscellaneous Business Services
7381 Detective, Guard, and Armored Car Services
7382 Security Systems Services
7383 News Syndicates
7384 Photofinishing Laboratories
7389 Business Services, Not Elsewhere Classified

Appendix B Excerpts from the Yahoo! classification system

Arts & Humanities
Literature, Photography...
Business & Economy
Companies, Finance, Jobs...
Computers & Internet
Internet, WWW, Software, Games...
Education
College and University, K-12...
Entertainment
Cool Links, Movies, Humor, Music...
Government
Elections, Military, Law, Taxes...
Health
Medicine, Diseases, Drugs, Fitness..
News & Media
Full Coverage, Newspapers, TV ...
Recreation & Sports
Sports, Travel, Autos, Outdoors ...
Reference
Libraries, Dictionaries, Quotations ...
Regional
Countries, Regions, US States...
Science
Animals, Astronomy, Engineering ...
Social Science
Archaeology, Economics, Languages...
Society & Culture
People, Environment, Religion ...

Expansion of Education: College and University
Academic Competitions@
Books@
College Entrance (440)
Colleges and Universities (10309)
Distance Learning@
Graduate Education (64)
Guidance (48)

Appendix C Excerpt from the *Encyclopedia Brittanica Propaedia*, 15th edition, 1992

Part One. Matter and Energy
Part Two: The earth
Part Three: Life on Earth
Part Four. Human Life
Part. Five Human Society
Part Six: Art
Part Seven: Technology
Part Eight: Religion
Part Nine: The History of Mankind
Part Ten. The Branches of Knowledge

Division I. Logic
10/11. History and Philosophy of Logic
10/12. Formal Logic, Metalogic, and Applied Logic
Division II. Mathematics
10/21. History and Foundations of Mathematics
10/22. Branches of Mathematics
10/23. Applications of Mathematics
Division III. Science
10/31. History and Philosophy of Science
10/32. The Physical Sciences
10/33. The Earth Sciences
10/34. The Biological Sciences
10/35. Medicine and Affiliated Disciplines
10/36. The Social Sciences and Psychology
10/37. The Technological Sciences
Division IV. History and the Humanities
10/41. Historiography and the Study of History
10/42. The Humanities and Humanistic Scholarship
Division V. Philosophy
10151. History of Philosophy
10/52. The Nature and the Divisions of Philosophy
10/53. Philosophical Schools and Doctrines

Breakdown of Section 10/36. The Social Sciences and Psychology and Linguistics

A. History of the social sciences
  1. Origins of the social sciences
    a. Precursors of the social sciences in the Middle Ages and the Renaissance
    b. Heritage of the Enlightenment: social reforms and revolution
  2. 19th-century developments in the social sciences
    a. The influence of new concepts in social, political, economic, and scientific theories
    b. Development of the separate disciplines; e.g. economics, political science, anthropology, sociology, social statistics, human geography
  3. 20th-century developments in the social sciences
    a. The influence of social upheaval in the non-Western world: the revolution of rising expectations
    b. The influence of Marxism
    c. The influence of Freudian ideas
    d. The changing character of the disciplines
      i. Specialization and cross-disciplinary approaches
      ii. The increasing professionalism of social scientists as consultants and decision makers in government and business
      iii. The introduction of mathematical and other quantitative methods: the use of computers
      iv. The influence of empiricism: the collection of data, the use of surveys and polls, the testing of theories
    e. Major theoretical influences: developmentalism, the social-systems approach, structuralism and functionalism
B. The nature of anthropology
  [See also Part Five, Division 11]
  1. The background of anthropology
  2. The scope and methods of anthropology: the division between cultural and physical anthropology
C. The nature of sociology
  [See also Part Five, Division III]
  1. The background of contemporary sociology
  2. The methodology of contemporary sociology
  3. The status of contemporary sociology
  4. Emergent trends in sociology
  5. Cognate disciplines: criminology, penology, social psychology, demography

Appendix D Extracts from the Library of Congress system

A GENERAL WORKS
B PHILOSOPHY. PSYCHOLOGY. RELIGION
C AUXILIARY SCIENCES OF HISTORY
D HISTORY: GENERAL AND OLD WORLD
E HISTORY: AMERICA
F HISTORY: AMERICA
G GEOGRAPHY. ANTHROPOLOGY. RECREATION
H SOCIAL SCIENCES
J POLITICAL SCIENCE
K LAW

1-7720	Law in general. Comparative and uniform law. Jurisprudence	
1-36.5	Periodicals	
37-44	Bibliography	
46	Monographic series	
48	Encyclopedias	
50-54	Dictionaries. Words and phrases	
58	Maxims. Quotations	
(64)	Yearbooks	
68-70	Directories	
85-89	Legal research	
94	Legal composition and draftsmanship	
100-103	Legal education	
109-110	Law societies. International bar associations	
115-130	The legal profession	
133	Legal aid. Legal assistance to the poor	
140-165	History of law	
170	Biography	
(175)	Congresses	
(176)-(177)	Collected works (nonserial)	
(179)	Addresses, essays, lectures	
181-184.7	Miscellany	
190-195	Ethnological jurisprudence. Primitive law	
201-487	Jurisprudence. Philosophy and theory of law	
	202	Periodicals
	212-213	Methodology
	215-218	History
	236	Universality and non-universality of law

S AGRICULTURE
T TECHNOLOGY
U MILITARY SCIENCE
V NAVAL SCIENCE
Z LIBRARY SCIENCE

A	General Works. 5th ed. (1998)
B-BJ	Philosophy. Psychology (1996)
BL, BM, BP, BQ	Religion: Religions. Hinduism, Judaism, Islam, Buddhism. 3rd ed. (1984)
BR-BV	Religion: Christianity, Bible (1987)
BX	Religion: Christian Denominations (1985)
C	Auxiliary Sciences of History (1996)
D-DJ	History (General), History of Europe, Part 1, 3rd ed. (1990)
DJK-DK	History of Eastern Europe: General, Soviet Union, Poland (1987)
DL-DR	History of Europe, Part 2, 3rd ed. (1990)
DS-DX	History of Asia, Africa, Australia, New Zealand, etc., (1998)
E-F	History: America (1995)
G	Geography. Maps. Anthropology. Recreation. 4th ed. (1976)
Subclass GE	Environmental Science (1976)
H	Social Sciences (1997)
J	Political Science (1997)
K	Law (General) 1998 Edition
KD	Law of the United Kingdom and Ireland, (1998)
KDZ	Law of the Americas, Latin America, and the West Indies (1984)
KG-KH	
KE	Law of Canada (1998)
KF	Law of the United States. (1999)
KJ-KKZ	Law of Europe (1989)
KJV-KJW	Law of France (1999)
KK-KKC	Law of Germany (1982)
KL-KWX	Law of Asia and Eurasia, Africa, Pacific Area and Antarctica, 1st ed. (1993)
KZ	Law of Nations, (1998)
L	Education. 1998 edition
m	Music and Books on Music. 1998 ed.
N	Fine Arts (1996)
P-PA	Philology and Linguistics (General). Greek Language and Literature. Latin Language and Literature (1997)
PB-PH	Modem European Languages (1999)
PG	Russian Literature (1948)
PJ-PK	Oriental Philology and Literature, Indo-Iranian Philology and Literature. 2nd ed. (1988)

PL-PM	Languages of Eastern Asia, Africa, Oceania; Hyperborean, Indian, and Artificial Languages. 2nd ed. (1988)
PN	Literature (General) (1997)
PR, PS, PZ	English and American Literature, Juvenile Belles Lettres. 1998 Edition.
PQ	French, Italian, Spanish, and Portuguese Literatures, (1998)
PT, Part 1	German Literature (1989)
PT, Part 2	Dutch and Scandinavian Literatures, 2nd ed. (1992)
Q	Science (1996)
R	Medicine (1995)
S	Agriculture (1996)
T	Technology (1995)
U-V	Military Science. Naval Science (1996)
Z	Bibliography and Library Science (1995)

Top levels of the Dewey Decimal System

000 Generalities
100 Philosophy & psychology
200 Religion
300 Social sciences
400 Language
500 Natural sciences & mathematics
600 Technology (Applied sciences)
700 The arts Fine and decorative arts
800 Literature & rhetoric
900 Geography & history

000 Generalities
   000 Generalities
      001 Knowledge
      002 The book
      003 Systems
      004 Data processing/ Computer science
      005 Computer programming, programs, data
      006 Special computer methods
      007 [unassigned]
      008 [unassigned]
      009 [unassigned]
   010 Bibliography
   020 Library & information sciences
   030 General encyclopedic works
   040 [unassigned]
   050 General serial publications
   060 General organizations & museology
   070 News media, journalism, publishing
   080 General collections
   090 Manuscripts & rare books

300 Social sciences
   310 Collections of general statistics
   320 Political science
   330 Economics
   340 Law

340 Law
341 International law
342 Constitutional & administrative law
343 Military, tax, trade, industrial law
344 Labor, social, education, cultural law
345 Criminal law
346 Private law
347 Civil procedure & courts
348 Law (Statutes), regulations, cases
349 Law of specific jurisdictions & areas
350 Public administration & military science
360 Social problems & services; association
370 Education
380 Commerce, communications, transportation
390 Customs, etiquette, folklore

600 Technology (Applied sciences)
610 Medical sciences Medicine
620 Engineering & allied operations
  620 Engineering & allied operations
  621 Applied physics
  622 Mining & related operations
  623 Military & nautical engineering
  624 Civil engineering
  625 Engineering of railroads & roads
  626 [unassigned]
  627 Hydraulic engineering
  628 Sanitary & municipal engineering
  629 Other branches of engineering
630 Agriculture & related technologies
640 Home economics & family living
650 Management & auxiliary services
660 Chemical engineering
670 Manufacturing
680 Manufacture for specific uses
690 Buildings

Extract of subject terms from Sears' list, 1986 edition

Annotations: Dewey Decimal Classification code; alternate term (x), target of "see also" cross reference (xx)

Inflation (Finance) 332.4
    See also Monetary policy; Paper money; Wageprice policy
    xx Finance; Monetary policy
Influenza 616.2
    x Flu; Grippe
    xx Cold (Disease)
Information centers. See Information services
Information, Freedom of. See Freedom of information
Information networks 001.5; 001.53; *004.6
    See also Computer networks; also types of information networks, e.g. Library information information networks; etc.
    x Automated networks; Networks,
Information networks-Continued Information
    xx Data transmission systems; Information services; Information storage and retrieval systems
Information science 020
    See also Documentation; Electronic data processing; Information services; Information storage and retrieval systems; Library science
    xx Communication
Information services 020
    See also

Archives Business	Information Libraries
services	Machine readable bibliographic data
Documentation	Reference services (Libraries)
Electronic publishing	Research
Hotlines (Telephone counseling)	Information networks
x Information centers	United Nations – Information
Information storage and retrieval systems	services

  .xx Documentation; Information science; Libraries; Research
Information storage and retrieval systems 025

Extract of subject terms from Library of Congress list

Annotated with the number of volumes in The University of Queensland Library indexed by each term.

Information Retrieval Study And Teaching	2
Information Retrieval Study And Teaching Elementary New Zealand Wellington Case Studies	1
Information Retrieval Study And Teaching Higher	1
Information Retrieval Study And Teaching Higher United States	1
Information Retrieval Study And Teaching Primary Australia	2
Information Retrieval Systems – see – Information Storage And Retrieval Systems	
Information Retrieval United States	1
Information Science – 7 Related Subjects	
Information Science	35
Information Science Abbreviations	5
Information Science Abstracts Periodicals	1
Information Science Acronyms	6
Information Science Australia	2
Information Science Australia Congresses	1
Information Science Australia Information Services Periodicals	1
Information Science Bibliography	1
Information Science Classification Books	1
Information Science Congresses	10
Information Science Dictionaries	13
Information Science Dictionaries Polyglot	1
Information Science Economic Aspects	1
Information Science Economic Aspects Bibliography	1
Information Science Government Policy – see – InformationPolicy	
Information Science Great Britain	1
Information Science Great Britain Periodicals	1
Information Science Handbooks Manuals Etc	1
Information Science Law And Legislation	1
Information Science Literature	
– see also – InformationScience Bibliography	
Information Science Mathematics Periodicals	1

Appendix I Selection from ACM keyword list[1]

General Terms These apply to any elements of the tree that are relevant.

Algorithms	Languages	Reliability
Design	Legal Aspects	Security
Documentation	Management	Standardization
Economics	Measurement	Theory
Experimentation	Performance	Verification
Human Factors		

Total number of terms is about 1500.

A.   General Literature (7)[2]
B.   Hardware (178)
C.   Computer Systems Organization (106)
D.   Software (236)
D.0   GENERAL (1)
D.1   PROGRAMMING TECHNIQUES (E)[3] (11)
D.2 SOFTWARE ENGINEERING (K.6.3) (100)
D.2.0 General (K.5.1) (3)
D.2.1 Requirements/Specifications (D.3.1) (5)
D.2.2 Design Tools and Techniques (REVISED) (14)
        Computer-aided software engineering (CASE)
        Decision tables
        Evolutionary prototyping (NEW)
        Flow charts
        Modules and interfaces
        Object-oriented design methods (NEW)
        Petri nets
        Programmer workbench**[4]

2 Approximate number of terms in this category.
3 See also the category indicated.
4 Indicates that the classification is no longer used as of January 1998, but that the item is still searchable for previously classified documents.

Extract of thesaurus terms from Sears' list, 1997 edition

Information science 020
BT Communication
NT Documentation
    Electronic data processing
    Information services
    Information systems
    Library science

Information services 025.5
UF Clearinghouses, Information
    Information centers
    Information clearinghouses

SA subjects, ethnic groups, classes of persons, individual persons, corporate bodies, industries, and names of countries, cities, etc., with the subdivision Information services, e.g. Business-Information services; United Nations-Information services; etc., to be added as needed

BT Information science
NT Archives
    Business-Information services
    Electronic publishing
    Hotlines (Telephone counseling)
    Information networks
    Information systems
    Machine readable bibliographic data
    Reference services (Libraries)
    United Nations-Information services
RT Documentation
    Libraries
    Research

Information systems
UF Computer-based information systems
    Data processing
    Data storage and retrieval systems

Information storage and retrieval systems [Former heading]
Punched card systems
BT   Bibliographic control
Bibliography
Computers
Documentation
Information science
Information services
NT   Database management
Electronic data processing
Expert systems (Computer science)

Information networks
Machine readable bibliographic data
Management information systems
Multimedia systems
Teletext systems
Videotex systems

RT   Libraries-Automation

Computer program languages
USE   Programming languages (Computers)

Computer programming
USE   Programming (Computers)

Computer programs
USE Computer software and subjects with the subdivision Computer programs, e.g.
Oceanography-Computer programs; to be added as needed

Computer software 005-3
UF   Computer programs [Former heading]
Programs, Computer
Software, Computer

SA types of computer software, e.g. Computer games; Electronic spreadsheets; etc.;
subjects with the subdivision Computer programs, e.g. Oceanography-Computer
programs; and names of individual computer programs, to be added as needed.

BT   Computer systems
NT   Computer assisted instruction-Authoring programs
Computer games
Computer software industry
Computer viruses
Database management – Computer programs
Electronic spreadsheets
Oceanography-Computer programs
Programming languages (Computers)
Utilities (Computer programs)
RT   Computers
Programming (Computers)

Computer software industry 338A
BT Computer software

Appendix K Extract of thesaurus terms from INSPEC, 1999 edition

Information science
More specific (narrower) terms:
  document deliver
  information analysis
  information centre
  information dissemination
  information nee
  information retrieval
  information retrieval systems
  information servic
  information storage
  information use
  vocabula
More general (broader) terms:
  computer applications
Related terms:
  information industry
  language translation
  libraries
  microforms
  publishing
  text editing

information services
More specific (narrower) terms:
  information networks
  information resources
More general (broader) terms:
  information science
Related terms:
  information analysis
  information centres
  information industry
  information needs
  libraries

  technical support services
  teletext
  viewdata

information systems
More specific (narrower) terms:
  database management systems
  digital libraries
  engineering information systems
  geographic information systems
  management information systems
  medical information systems
  personal information systems
  public information systems
  scientific information systems
  traffic information systems
More general (broader) terms
computer applications
Related terms:
  information retrieval systems
  information technology
  strategic planning
  systems re-engineering

database management systems
More specific (narrower) terms:
  active databases
  database indexing
  database machines
  deductive databases
  distributed databases
  meta dataprogramming
  multimedia databases
  object-oriented databases
  relational databases

statistical databases
temporal databases
very large databases
visual databases
More general (broader) terms:
  file organisation
  information systems
Related terms:
  application generators
  concurrency control
  data integrity
  data models
  database theory
  decision support systems
  geographic information systems
  group decision support systems
  hypermedia
  integrated software
  multimedia systems
  query languages
  query processing
  systems re-engineering
  transaction processing

computer software
More specific (narrower) terms:
  :automatic test software
  complete computer programs
  computer communications software
  expert system shells
  integrated software
  macros
  public domain software
  software packages
software standards
  subroutines
  systems software
Related terms:
  data description
  firmware
  programming languages
  programming theory
  software engineering

Appendix L **Images of spikes**

Towards the end of the 19th century, it became fashionable in Australia, and doubtless other countries, for houses to have fences made of iron rods arranged vertically. These rods were typically topped with spikes. In 1990, I lived in a neighbourhood in Sydney which had been developed at that time. I spent the summer holiday traversing the streets in the surrounding neighbourhoods photographing fence spikes.

Our house had a fence with spike 1, but there were a great variety of spikes, the most common being a fleur-de-lis (spike 2). Many fences had larger spikes above the main posts (e.g. spikes 62 and 65), and some had also smaller spikes atop smaller filler bars (e.g. 20). Below is a selection from that collection of photographs.

One problem I encountered in making the collection was telling whether an unusual spike I encountered was already in the collection. It would have been useful to have carried the collection in a palm computer and to have had some way of indexing it so I could quickly determine whether a new spike was already there.

42

44

45

55

56

57

58

60

62

64

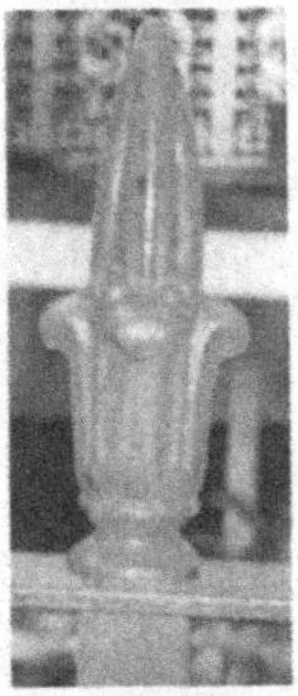

65

66

71

75

76

79

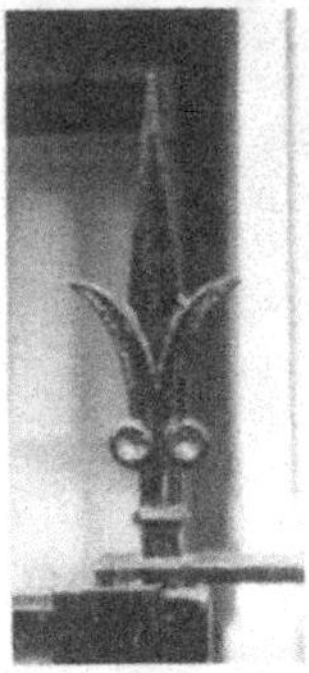

81 82 89 101 102 103 104 105 106 107 108 109

Bibliography

Abbott, Edwin Abbott (1952) *Flatland: A Romance of Many Dimensions*. New York: Dover.

Ayres, Leonard P. (1919) *The War with Germany*. Washington DC: US Government.

Bates, M.J. (1986) "Subject access in online catalogs: a design model", *JASIS*, Vol. 37, No. 6, pp. 357–76.

Bates, M.J. (1989) "The design of browsing and berrypicking techniques for the online search interface", *Online Review*, Vol. 13, No. 12, pp. 29–38.

Benedikt, Michael (ed.) (1991a) *Cyberspace: First Steps*. Cambridge, MA: MIT Press.

Benedikt, Michael (1991b) "Cyberspace: some proposals", Chapter 7 in Benedikt (1991a).

Chernoff, Herman (1973) "The use of faces to represent points in k-dimensional space graphically", *Journal of the American Statistical Association*, Vol. 68, pp. 361–68.

Cifuentes, Cristina and Fitzgerald, Anne (1996) "Copyrighting shareware on the Internet", *Computer*, Vol. 26, No. 3, pp. 110–111.

Colomb, R.M. (2001) "Why do people pay for information?", *Prometheus*, Vol. 19, No. 1, pp. 45–53.

de Certeau, Michel (1984) *The Practice of Everyday Life*. Berkeley: University of California Press, Chapter 12 "Reading as Poaching".

Dodge, Martin and Kitchin, Rob (2001) *Atlas of Cyberspace*. San Francisco: Morgan Kaufmann.

Ellis, J. (ed.) (1993) *Keeping Archives* (2nd edn). Port Melbourne: D.W. Thorpe.

Elmasri, R. and Navathe, S.B. (2000) *Fundamentals of Database Systems* (3rd edn). Reading, MA: Addison Wesley.

Encyclopedia Brittanica, Propaedia.

Fellbaum, Christiane (ed.) (1998) *Wordnet: An Electronic Lexical Database (Language, Speech and Communication)*. Cambridge, MA: MIT Press.

Foucault, Michel (1986) *The History of Sexuality Vol. 1*. New York: Vintage Books, Part 3, Part 4, Chapter 2.

Foucault, Michel (1970) *The Order of Things: An Archaeology of the Human Sciences*. London: Tavistock Publications, Chapters 2, 3, 5.

Gilbert, E.W. (1958) "Pioneer maps of health and disease in England", *Geographical Journal*, Vol. 124, pp. 172–83.

Goldfarb, Charles F. and Prescod, Paul (1998) *XML Handbook*. Englewood Cliffs: Prentice Hall

Gould, Stephen Jay (1977) *Ever since Darwin: Reflections in Natural History*. New York: Norton, Chapter 29.

Gould, Stephen Jay (1983) *Hen's Teeth and Horse's Toes*. New York/London: Norton, Chapter 5.

Gould, Stephen Jay (1985) *The Flamingo's Smile: Reflections in Natural History*. New York: Norton, Chapters 10, 11.

Gould, Stephen Jay (ed.) (1993) *The Book of Life*. London: Ebury Hutchinson.

Johnson, Richard F. and Selander, Robert K. (1964) "House sparrows: rapid evolution of races in North America", *Science*, Vol. 144, 1 May, pp. 548–50.

Kinsey, Alfred C., Pomeroy, Wardell B. and Martin, Clyde E. (1948) *Sexual Behaviour in the Human Male*. Philadelphia: Saunders.

Korfhage, Robert R. (1997) *Information Storage and Retrieval*. New York: Wiley.

Lynch, Clifford A. (1995) "Networked information resource discovery: an overview of current issues", *IEEE Journal on Selected Areas of Communications*, Vol. 13, No. 8, pp 1505–22.

Macquarie Dictionary (1987) Sydney: Macquarie Library.

Megginson, David (1998) *Structuring XML Documents*. Englewood Cliffs: Prentice Hall.

Nation, David A., Plaisant, Catherine, Marchionini, Gary and Komlodi, Anita (1997). *Visualizing websites using a hierarchical table of contents browser: WebTOC* ftp://ftp.cs.umd.edu/pub/hcil/Demos/WebTOC/Paper/WebTOC.html.

Plato, *The Phaedrus*. <http://classics.mit.edu/Plato/phaedrus.html>

Reka, Albert, Jeong, Hawoong and Barabasi, Albert-Laszlo (1999) "Diameter of the World-Wide Web", *Nature*, Vol. 401, 9 September, p. 130.

Rowley, Jennifer E. (1992) *Organizing Knowledge* (2nd edn). Aldershot: Gower.

Salton, Gerard (1989) *Automatic Text Processing: The Transformation, Analysis, and Retrieval of Information by Computer*. Reading, MA: Addison-Wesley.

Surgeon-General (1964) Report of the Advisory Committee to the Surgeon General [of the USA], *Smoking and Health*, Washington DC.

Tufte, Edward R. (1983) *The Visual Display of Quantitative information*. Cheshire, CN: Graphics Press.

Tufte, Edward R. (1990) *Envisioning Information*. Cheshire, CN: Graphics Press.

Tufte, Edward R. (1997) *Visual Explanations: Images and Quantities, Evidence and Narrative*. Cheshire, CN: Graphics Press.

Van Rijsbergen, C.J. (1979) *Information Retrieval*. London: Butterworths.

Weber, Robert Philip (1990) *Basic Content Analysis* (2nd edn). Newbury Park, CA: Sage Publications.

Wexelblat, Alan (1991) "Giving meaning to place: semantic spaces", Chapter 9 in Benedikt (1991a).

Wittgenstein, Ludwig (1953) *Philosophical Investigations*. Oxford: Blackwell.

Index

Glossary entries for key terms are in bold. Definitions in context are in Key Terms sections indicated by bold page number.